EW 101

A First Course in Electronic Warfare

For a listing of recent titles in the *Artech House Radar Library*, turn to the back of this book.

EW 101

A First Course in Electronic Warfare

David Adamy

Artech House
Boston • London
www.artechhouse.com

Library of Congress Cataloging-in-Publication Data
Adamy, David.
    EW 101: a first course in electronic warfare / David L. Adamy.
        p. cm. — (Artech House radar library)
    Includes bibliographical references and index.
    ISBN 1-58053-169-5 (alk. paper)
        1. Electronics in military engineering. I. Title. II. Series.

    UG485 .A33   2000   00-048095
    623'.043—dc21   CIP

British Library Cataloguing in Publication Data
Adamy, David
    EW 101: a first course in electronic warfare. — (Artech
    House radar library)
    1. Electronics in military engineering
    I. Title
    623'.043

    ISBN 13 978-1-58053-169-6

Cover design by Igor Valdman

© 2001 ARTECH HOUSE, INC.
685 Canton Street
Norwood, MA 02062

International Standard Book Number: 1-58053-169-5
Library of Congress Catalog Card Number: 00-048095

20 19 18 17 16 15 14 13 12 11

This book is dedicated to my colleagues in the EW profession, both in and out of uniform. Some of you have gone repeatedly into harm's way, and most of you have often worked long into the night to do things that are beyond the comprehension of the average person. Ours is a strange and challenging profession, but most of us can't imagine following any other.

Contents

9   Jamming   177

Preface

EW 101 has been a popular column in the *Journal of Electronic Defense* (JED) for several years—covering various aspects of electronic warfare (EW) in two-page bites on a month-to-month basis. No one really knows why a particular column is popular, but the suspicion is that it hits a level that is useful to a wide range of people. This book includes the material in the first 60 columns of the series—organized into chapters with some extra material for continuity.

The target audiences for this book are: new EW professionals, specialists in some part of EW, and specialists in technical areas peripheral to EW. Another target group is managers who used to be engineers, who now must make decisions based on input from others (who may or may not be trying to break the laws of physics). In general, the book is intended for those to whom a general overview, a grasp of the fundamentals, and the ability to make general level calculations are valuable. Finally, it is an answer to the many individuals who have been trying to collect complete sets of EW 101 columns (some of which are hard to find).

As this book goes to press, the EW 101 columns continue in JED. EW is a broad field, so there is plenty to talk about for several more years. (Look for a series of editions of this book in the distant future.)

I sincerely hope that this book is useful to you on the job; that it saves you time and worry; and that it may even keep you out of trouble from time to time.

1

Introduction

Like the monthly tutorial articles from which it sprang, this book is intended to provide a top-level view of the broad, important, and fascinating field of electronic warfare (EW). Here are some generalities about the book:

- It is not intended for experts in the field, although it is hoped that experts in other fields and experts in subfields within electronic warfare may find it useful.
- It is intended to be easy to read. Technical material (contrary to popular opinion) does not need to be boring to be useful.
- It is true to the columns in level, style, and content. The material, however, has been reorganized to be useful as a book. Most of the columns are ordered into a logical sequence of chapters. Where two or more subject areas were covered in the same column, the material has been moved to its respective appropriate chapters.

The coverage of technical material in this book is intended to be accurate as opposed to precise. The formulas, in most cases, are accurate to 1 dB—which is accurate enough for most system level design work. Even when much greater precision is required, almost all old-hand systems engineers run the basic equations to 1 dB first, then turn loose the computer experts to drive to the required precision. The problem with highly precise mathematics is that you can get lost down in the details and make mistakes of orders of magnitude. These mistakes are sometimes incorrect assumptions or (more often) an incorrectly stated problem. Order of magnitude errors get you (and probably your boss) into big trouble; they are worth avoiding.

1

When you work the problem to 1 dB, using the simple dB form equations in this book, you will quickly derive a set of approximate answers. Then, you can sit back and ask yourself if the answers make sense. Compare the results to the results of other, similar problems . . . or just apply common sense. At this point, it is easy to revisit the assumptions or clarify the statement of the problem. Then, when you apply the considerable facilities, staff time, money, and (perhaps) stomach acid required to complete the detailed calculations, they have an even chance of coming out right the first (or nearly the first) time.

The Scope of the Book

This book covers most of the radio frequency (RF) aspects of electronic warfare at the system level. This means that it talks more about what hardware and software do than specifically how they do what they do. It avoids complex mathematics, assumes that the reader has algebra and some trigonometry skills, and diligently avoids calculus.

More Detailed Information

There are many recommended sources of more detailed EW information. These include textbooks, professional journals, and technical magazines. The following is far from a complete list of the reference material available. It is, however, a reasonable starting list, and it is the set of references that supported the preparation of the EW 101 articles. Some of these references are complex and others are fairly easy for beginners in the field (and for those of us who have been out of school for a long time), but they all contain solid, useful information relevant to EW.

General electronic warfare texts:
- *Electronic Warfare*, D. Curtis Schleher (Artech House)
- *Introduction to Electronic Defense Systems*, Filippo Neri (Artech House)
- *Electronic Warfare*, David Hoisington (Lynx)
- *Applied ECM* (three volumes), Leroy Van Brunt (EW Engineering, Inc.)

Books on more specific EW subjects:
- *Radar Vulnerability to Jamming*, Robert Lothes, Michael Szymanski, and Richard Wiley (Artech House)

- *Electronic Intelligence: The Interception of Radar Signals,* Richard Wiley (Artech House)
- *Radar Cross Section,* Eugene Knott, John Shaeffer, and Michael Tuley (Artech House)
- *Introduction to Radar Systems,* Merrill Skolnik (McGraw-Hill)
- *Introduction to Airborne Radar,* George Stimson (SciTech)

Books on new modulations:
- *Spread Spectrum Communications Handbook,* Marvin Simon et al. (McGraw-Hill)
- *Detectability of Spread-Spectrum Signals,* Robin and George Dillard (Artech House)
- *Spread Spectrum Systems with Commercial Applications,* Robert Dixon (Wiley)
- *Principles of Secure Communication Systems,* Donald Torrieri (Artech House)

EW handbooks:
- *International Countermeasures Handbook* (Horizon House)
- *EW Handbook (Journal of Electronic Defense)*

Magazines with articles related to EW:
- *Journal of Electronic Defense*
- *IEEE Transactions on Aerospace and Electronic Systems* (from IEEE AES working group)
- *Signal Magazine*
- *Microwave Journal*
- *Microwaves*

Generalities About EW

Electronic warfare is defined as the art and science of preserving the use of the electromagnetic spectrum for friendly use while denying its use to the enemy. The electromagnetic spectrum is, of course, from DC to light (and beyond). Thus, electronic warfare covers the full radio frequency spectrum, the infrared spectrum, the optical spectrum, and the ultraviolet spectrum.

As shown in Figure 1.1, EW has classically been divided into:

- Electromagnetic support measures (ESM)—the receiving part of EW;
- Electromagnetic countermeasures (ECM)—jamming, chaff, and flares used to interfere with the operation of radars, military communication, and heat-seeking weapons;
- Electromagnetic counter-countermeasures (ECCM)—measures taken in the design or operation of radars or communication systems to counter the effects of ECM.

Antiradiation weapons (ARW) and directed-energy weapons (DEW) were not considered part of EW, even though it was well understood that they were closely allied with EW. They were differentiated as weapons.

In the last few years, the subdivisions of the EW field have been redefined as shown in Figure 1.2 in many (but not all) countries. Now the accepted definitions (in NATO) are:

- Electronic warfare support (ES)—which is the old ESM;
- Electronic attack (EA)—which includes the old ECM (jamming, chaff, and flares), but also includes antiradiation weapons and directed-energy weapons;
- Electronic protection (EP)—which is the old ECCM.

ESM (or ES) is differentiated from signal intelligence (SIGINT)—which comprises communications intelligence (COMINT) and electronic intelligence (ELINT)—even though all of these fields involve the receiving of enemy transmissions. The differences, which are becoming increasingly vague as the complexity of signals increases, are in the purposes for which transmissions are received.

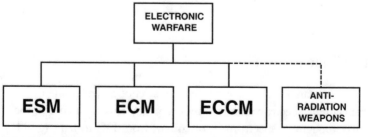

Figure 1.1   Electronic warfare classically has been divided into ESM, ECM, and ECCM. Antiradiation weapons were not part of EW.

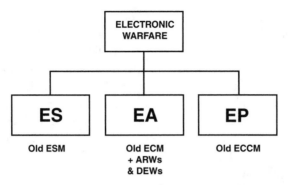

Figure 1.2   Current NATO electronic warfare definitions divide EW into ES, EA, and EP. EA now includes antiradiation and directed-energy weapons.

- COMINT receives enemy communications signals for the purpose of extracting intelligence from the information carried by those signals.

- ELINT receives enemy noncommunication signals for the purpose of determining the details of the enemy's electromagnetic systems so that countermeasures can be developed. Thus, ELINT systems normally collect lots of data over a long period of time to support detailed analysis.

- ESM/ES, on the other hand, collects enemy signals (either communication or noncommunication) with the object of immediately doing something about the signals or the weapons associated with those signals. The received signal might be jammed or its information handed off to a lethal response capability. The received signals can also be used for situation awareness; that is, identifying the types and location of the enemy's forces, weapons, or electronic capability. ESM/ES typically gathers lots of signal data to support less extensive processing with a high throughput rate. ESM/ES typically determines only *which* of the known emitter types is present and where they are located.

How to Understand Electronic Warfare

It is the contention of the author that the key to understanding EW principles (particularly the RF part) is to have a really good understanding of radio propagation theory. If you understand how radio signals propagate, there is a logical progression to understanding how they are intercepted, jammed, or protected. Without that understanding, it seems (to the author) that it is almost impossible to really get your arms around EW.

Once you know a few simple formulas, like the one-way link equation and the radar range equation in their dB forms, you will most likely be able to run EW problems in your head (to 1-dB accuracy). If you get to that point, you can quickly "cut to the chase" when facing an EW problem. You can quickly and easily check to see if someone is trying to break the laws of physics. (OK, a piece of scratch paper is allowed as long as it is crudely torn from a pad—your colleagues will still class you as an EW expert when you get them out of trouble.)

On to the Specifics

- Chapter 2 covers some background math: dBs, the link equations, and spherical trig (in case you forgot).
- Chapter 3 covers antennas: types, definitions, and parametric trade-offs.
- Chapter 4 covers receivers: types, definitions, applications, and sensitivity calculations.
- Chapter 5 covers EW processing: signal identification, control mechanisms, and operator interface.
- Chapter 6 covers search techniques, limitations, and tradeoffs.
- Chapter 7 covers low probability of intercept signals, with the primary focus on LPI communications.
- Chapter 8 covers all of the common emitter location techniques used in EW systems.
- Chapter 9 covers jamming: concepts, definitions, limitations, and equations.
- Chapter 10 covers radar decoys: active and passive, including appropriate calculations.
- Chapter 11 covers simulation for concept evaluation, training, and system testing.

For those who have a favorite EW 101 column, there is a column-to-chapter cross-reference in Appendix A at the end of the book for your convenience.

2

Basic Mathematical Concepts

This chapter covers the basic math that underlies the EW concepts covered in the other chapters. It includes a discussion of the dB form of numbers and equations, radio propagation, and spherical trigonometry.

2.1 dB Values and Equations

In any professional activity that includes consideration of radio propagation, signal strength, gains, and losses are often stated in dB form. This allows the use of dB forms of equations that are typically easier to use than the original forms.

Any number expressed in dB is logarithmic, which makes it convenient to compare values that may differ by many orders of magnitude. For convenience, we will call numbers in non-dB form "linear" to differentiate them from the logarithmic dB numbers. Numbers in dB form also have the great charm of being easy to manipulate:

- To multiply linear numbers, you add their logarithms.
- To divide linear numbers, you subtract their logarithms.
- To raise a linear number to the nth power, you multiply its logarithm by n.
- To take the nth root of a linear number, you divide its logarithm by n.

To take maximum advantage of this convenience, we put numbers in dB form as early in the process as possible, and we convert them back to linear forms as late as possible (if at all). In many cases, the most commonly used forms of answers remain in dB.

It is important to understand that any value expressed in dB units must be a ratio (which has been converted to logarithmic form). Common examples are amplifier or antenna gain and losses in circuits or in radio propagation.

2.1.1 Conversion to and from dB Form

A linear number (N) is converted to dB form by the formula:

$$N(\text{dB}) = 10 \log_{10}(N)$$

For most of the equations in this book, we just say 10 log (N), with the logarithm to the base 10 understood. To do this operation using a scientific calculator, enter the linear form number, then touch the "log" key, then multiply by 10.

dB values are converted to linear form with the equation:

$$N = 10^{N(\text{dB})/10}$$

Using a scientific calculator, enter the dB form number, divide it by 10, then touch the second function key, then the log key. This process is also described as taking the "antilog" of the dB value divided by 10.

For example, if an amplifier has a gain factor of 100, we can say it has 20-dB gain, because:

$$10 \log (100) = 10 \times 2 = 20 \text{ dB}$$

Reversing the process to find the linear form gain of a 20-dB amplifier:

$$10^{20/10} = 100$$

2.1.2 Absolute Values in dB Form

To express absolute values as dB numbers, we first convert the value to a ratio with some understood constant value. The most common example is signal strength expressed in dBm. To convert a power level to dBm, we divide it by 1 milliwatt and then convert to dB form. For example, 4 watts equals 4000 milliwatts. Then convert 4000 to dB form to become 36 dBm. The small "m" indicates that this is a ratio to a milliwatt.

$$10 \log (4000) = 10 \times 3.6 = 36 \text{ dBm}$$

Then, to covert back to watts:

$$\text{Antilog }(36/10) = 4000 \text{ milliwatts} = 4 \text{ watts}$$

Other examples of dB forms of absolute values are shown in Table 2.1.

Table 2.1
Common dB Definitions

dBm	= dB value of Power / 1 milliwatt	Used to describe signal strength
dBW	= dB value of Power / 1 watt	Used to describe signal strength
dBsm	= dB value of Area / 1 meter^2	Used to describe antenna area or radar cross-section
dBi	= dB value of antenna gain relative to the gain of an isotropic antenna	0 dBi is, by definition, the gain of an omnidirectional (isotropic) antenna

2.1.3 dB Equations

In this book, we use a number of dB forms of equations for convenience. These equations have one of the following forms, but can have any number of terms:

$$A(\text{dBm}) \pm B(\text{dB}) = C(\text{dBm})$$

$$A(\text{dBm}) - B(\text{dBm}) = C(\text{dB})$$

$$A(\text{dB}) = B(\text{dB}) \pm N \log (\text{number not in dB})$$

where N is a multiple of 10.

This last equation form is used when the square (or higher order) of a number is to be multiplied. An important example of this last formula type is the equation for the spreading loss in radio propagation:

$$L_S = 32 + 20 \log (d) + 20 \log (f)$$

where L_S = spreading loss (in dB); d = link distance (in km); and f = transmission frequency (in MHz).

The factor 32 is a fudge factor that is added to make the answer come out in the desired units from the most convenient input units. It is actually

4π squared, divided by the speed of light squared, multiplied and divided by some unit conversion factors—and the whole thing converted to dB form and rounded to a whole number. The important thing to understand about this fudge factor (and the equation that contains it) is that it is correct only if exactly the correct units are used. The distance must be in kilometers and the frequency must be in megahertz—otherwise, the loss value will not be correct.

2.2 The Link Equation for All EW Functions

The operation of every type of radar, military communication, signals intelligence, and jamming system can be analyzed in terms of individual communication links. A link includes one radiation source, one receiving device, and everything that happens to the electromagnetic energy as it passes from the source to the receiver. The sources and the receivers can take many forms. For example, when a radar pulse reflects from the skin of an aircraft, the reflecting mechanism can be treated like a transmitter. Once the reflected pulse leaves the aircraft's skin, it obeys the same laws of propagation that apply to a communication signal from one push-to-talk tactical transceiver to another.

2.2.1 The "One-Way Link"

The basic communication link, or "one-way link" as it is sometimes called, consists of a transmitter (XMTR), a receiver (RCVR), transmitting and receiving antennas, and a propagation path between the two antennas. Figure 2.1 shows what happens to the strength of the radio signal as it passes through the link. This diagram shows signal strength in dBm and increases and decreases of signal strength in dB.

As drawn, Figure 2.1 is for a line-of-sight link (i.e., the transmitting and receiving antennas can "see" each other and the transmission path between the two doesn't get too close to land or water) in good weather, which we will consider first. Later, we'll add the effects of bad weather and non-line-of-sight propagation to our link calculations. The signal leaves the transmitter at some power level in dBm. It is "increased" by the transmitting antenna gain. (If the antenna gain is less than unity, or 0 dB, the signal strength leaving the antenna is less than the transmitter output power.) The signal power leaving the antenna is called the effective radiated power (ERP) and is commonly stated in dBm. The radiated signal is then attenuated

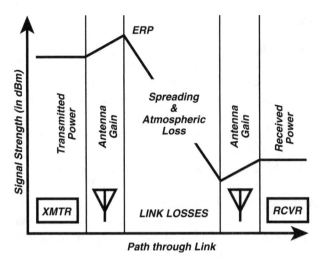

Figure 2.1 To calculate the received signal level (in dBm), add the transmitting antenna gain (in dB), subtract the link losses (in dB), and add the receiving antenna gain (in dB) to the transmitter power (in dBm).

by various factors as it propagates between the transmitting and receiving antennas. For a line-of-sight link in good weather, the attenuation factors are just the spreading loss and the atmospheric loss. The signal is "increased" by the receiving antenna gain (which can be either a positive or negative number, depending on the nature of the antenna). The signal then reaches the receiver at the "received power."

The process described by Figure 2.1 is known as the "link equation" or the "dB form of the link equation." Although spoken of in the singular, the "link equation" actually refers to a set of several equations by which we can calculate the signal strength at any point in the propagation process in terms of all of the other elements.

A typical example of a link equation application is:

Transmitter Power (1W) = +30 dBm
Transmitting Antenna Gain = +10 dB
Spreading Loss = 100 dB
Atmospheric Loss = 2 dB
Receiving Antenna Gain = +3 dB

Received Power = +30 dBm + 10 dB − 100 dB − 2 dB + 3 dB
= −59 dBm

2.2.2 Propagation Losses

The two challenging elements of the above equation are the spreading loss (also called the "space loss") and the atmospheric loss. (No offense to the transmitter and antenna manufacturers, but we just read the specification sheets to determine their parameters, while we have to calculate the propagation losses for every different situation.) Both of these propagation loss factors vary with the propagation distance and the transmitting frequency. First, we'll get spreading loss the easy way, from the nomograph in Figure 2.2. To use this chart, draw a line from the frequency on the left-hand scale (1 GHz in the example) to the transmission distance on the right-hand scale (20 km in the example). The line crosses the middle scale at 119 dB, the value of the spreading loss for that frequency and that transmission distance. There is also a simple dB formula which calculates spreading loss:

$$L_S \ (in \ dB) = 32.4 + 20 \log_{10}(distance \ in \ km) + 20 \log_{10}(frequency \ in \ MHz)$$

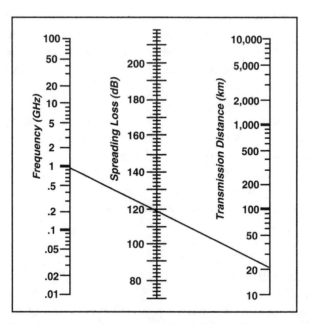

Figure 2.2 Spreading loss can be determined by drawing a line from the frequency (in GHz) to the transmission distance (in km) and reading the spreading loss (in dB) on the center scale.

Remember that this is for line-of-sight propagation in good weather. The factor 32.4 combines all of the unit conversions to make the answer work out—it is only valid for distance in kilometers and frequency in megahertz. The 32.4 factor is typically rounded to 32 when the link equation is used to 1-dB accuracy.

One other point about spreading loss: The loss value derived from the nomograph and from the above formula is for the spreading loss between two isotropic antennas (that is, antennas with "unity" or 0-dB gain). This makes the bookkeeping easy, because we then add the antenna gains to the equation as independent numbers. The formula (which is also the basis for the nomograph) is derived from the fact that a truly isotropic transmitting antenna radiates its energy spherically, so the effective radiated power (ERP) is evenly distributed over the surface of an expanding sphere. An isotropic receiving antenna has an "effective area" which is a function of frequency. The effective area of the (isotropic) receiving antenna determines the amount of the surface of the sphere (which has a radius equal to the distance from the transmitter to the receiver) that a unity gain antenna will collect. The spreading loss formula is the ratio of the whole surface area of the sphere to the area of the isotropic antenna (at the operating frequency). The derivation is "left as an exercise for the reader" if you are really hard up for something to do.

Atmospheric attenuation is nonlinear, so it is best handled by just reading the values from Figure 2.3. In the example, the transmission frequency is 50 GHz. Draw a line from 50 GHz up to the curve and then straight left to determine the atmospheric loss per kilometer of transmission path length. In the example, it is 0.4 dB per kilometer, so a 50-GHz signal traveling 20 km would have 8 dB of atmospheric attenuation. Please note that at the frequencies used for most point-to-point tactical communications, the atmospheric attenuation is quite low and is often ignored in link calculations. However, it becomes very significant at high microwave and millimeter-wave frequencies and in transmission through the whole atmosphere to and from Earth satellites.

2.2.3 Receiver Sensitivity

Although receiver sensitivity will be discussed in detail in Chapter 4, it should be understood for the present discussion that the sensitivity of a receiver is defined as the smallest signal (i.e., the lowest signal strength) that it can receive and still provide the proper specified output.

If the received power level is at least equal to the receiver sensitivity, communication takes place over the link. For example, if the received power

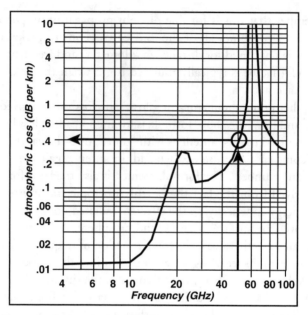

Figure 2.3   Atmospheric attenuation (in dB) per kilometer of transmission path can be determined by drawing a line from the frequency (in GHz) up to the curve and then left to the attenuation scale.

is −59 dBm (as in the above example) and the receiver sensitivity is −65 dBm, communication will take place. Because the received signal is 6 dB higher than the receiver's sensitivity specification, we say that the link has 6 dB of margin.

2.2.4 Effective Range

At the maximum link range, the received power will be equal to the receiver sensitivity. Thus, we can set the received power equal to the sensitivity and solve for the distance. For simplicity, let's work an example at 100 MHz where the atmospheric loss is negligible over normal terrestrial link distances.

The transmitter power is 10 watts (which equals +40 dBm), the frequency is 100 MHz, the transmitting antenna gain is 10 dB, the receiving antenna gain is 3 dB, and the receiver sensitivity is −65 dBm. There is line of sight between the two antennas. What is the maximum link range?

$$P_R = P_T + G_T - 32.4 - 20\log(f) - 20\log(d) + G_R$$

where P_R = received power (in dBm); P_T = transmitter output power (in dBm); G_T = transmitting antenna gain (in dB); f = transmitted frequency (in MHz); d = transmission distance (in km); G_R = receiving antenna gain (in dB).

Setting P_R = Sens (the receiver sensitivity) and solving for 20 log(d):

$$\text{Sens} = P_T + G_T - 32.4 - 20 \log(f) - 20 \log(d) + G_R$$

$$20 \log(d) = P_T + G_T - 32.4 - 20 \log(f) + G_R - \text{Sens}$$

Plugging in the above stated dB values:

$$20 \log(d) = +40 + 10 - 32.4 - 20 \log(100) + 3 - (-65) =$$
$$+40 + 10 - 32.4 - 40 + 3 + 65 = 45.6$$

Then, solving for d, the effective range is found to be:

$$d = \text{Antilog} (20 \log(d)/20) = \text{Antilog} (45.6/20) = 191 \text{ km}$$

2.3 Link Issues in Practical EW Applications

The basic link equation takes many forms in the various EW systems and engagements. There is also an important artifice that greatly simplifies our understanding of what is going on in EW links.

2.3.1 Power Out in the Ether Waves

The link equation formulas presented in this book (and used by most people who practice system-level EW magic for a living) contain a serious logical flaw. But they make our lives so much simpler that we are ready to fight to defend them against those to whom rigorousness is a theological issue. The flaw is that we state the power of signals "out in the ether waves"—that is, between a transmitting antenna and a receiving antenna—in dBm. The problem is that dBm is just the logarithmic representation of milliwatts. Signal strength in dBm is "power," and electrical power is only defined inside a wire or a circuit. While propagating from a transmitting to a receiving antenna, signals must be accurately described in terms of their "electric intensity," which is most commonly quantified in microvolts per meter (µV/m). (See Figures 2.4 and 2.5.)

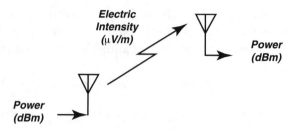

Figure 2.4   It is actually only rigorous to speak of signal strength in dBm inside a wire or a circuit. Out in the "ether waves" it is correct to speak of electric intensity in μV/m.

So how do we come up with dBm values for propagating waves that produce correct answers when applied to the analysis of links? We use an *artifice* [*n.* **1.** An ingenious stratagem; maneuver. **2.** Subtle or deceptive craft; trickery]. The artifice creates an imaginary, ideal unity-gain antenna located at the point in space where we want to assign a signal strength to the signal of interest. That signal strength (in dBm) would be present in the output of the imaginary antenna. Thus, the effective radiated power (ERP) would be output by the imaginary antenna if it were located on a line from the transmitting antenna to the receiving antenna and almost touching the transmitting antenna (ignoring near field effects, of course). Likewise, in a representation of the power arriving at the receiving antenna (often called P_A), the imaginary antenna would be on the same line, but almost touching the receiving antenna.

2.3.2   Sensitivity in μV/m

Receiver sensitivity is sometimes stated in μV/m rather than in dBm. This is particularly true for devices in which an intimate and complex relationship

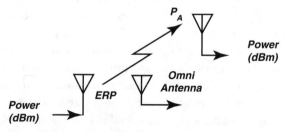

Figure 2.5   A radiated signal is often described in terms of what an ideal receiver and omnidirectional antenna would receive.

between the antenna(s) and the receiver exists. The best example is probably a direction-finding system with a space-diverse antenna array. Fortunately, a pair of simple dB-type formulas (based on that imaginary unity-gain antenna) translate between μV/m and dBm. In all of the equations in this chapter, "log" means log to the base 10. To convert from μV/m to dBm:

$$P = -77 + 20 \log(E) - 20 \log(F)$$

where P = signal strength (in dBm); E = electric intensity (in μV/m); F = frequency (in MHz).

To convert from dBm to μV/m:

$$E = 10^{(P + 77 + 20 \log [F])/20}$$

These formulas are based on the equations:

$$P = (E^2 A)/Z_0$$

and

$$A = (G c^2)/(4\pi F^2)$$

where P = signal strength (in W); E = electric intensity (in V/m); A = antenna area (in m²); Z_0 = impedance of free space (120π ohms); G = antenna gain (= 1 for isotropic antenna); c = speed of light (3×10^8 m/sec); and F = frequency (in Hz).

You are welcome to derive these, if that's your idea of a good time. (It's really quite straightforward if you remember the unit conversion factors and then convert the whole, combined equation to dB form.)

2.3.3   "Links" in Radar Operation

Many textbooks present the radar range equation in the form most useful to radar people, since the equation focuses on how well the radar is doing its job. However, for EW people it is more useful to consider the radar range equation in terms of its component "links," as shown in Figure 2.6, and to handle everything in terms of dB and dBm. This allows us to deal with the radar power arriving at a target; the power we must generate with a jammer if we are to equal (or exceed by some fixed factor) the power returned by that target to the radar receiver; and many other useful values.

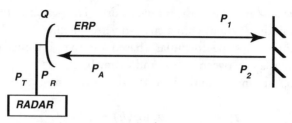

Figure 2.6 For convenience in EW applications, the radar range equation can be described as a series of links.

You will recognize the expression for spreading loss presented earlier $[32.4 + 20 \log(D) + 20 \log(F)]$, but for convenience the 32.4 factor is normally rounded to 32. There is also a handy expression for the signal reflection factor caused by the radar cross-section of the target $[-39 + 10 \log(\sigma) + 20 \log(F)]$. This expression will be derived and treated in much more detail in Chapter 10.

P_T is the radar's transmitter power into its antenna (dBm). G is the main beam gain of the radar antenna (dB). ERP is the effective radiated power. P_1 is the signal power arriving at the target (dBm). P_2 is the signal power reflected from the target back toward the radar (dBm). P_A is the signal power arriving at the radar's antenna (dBm). P_R is the received power (into the radar receiver) (dBm).

In dB form:

$$ERP = P_T + G$$
$$P_1 = ERP - 32 - 20 \log(D) - 20 \log(F)$$
$$= P_T + G - 32 - 20 \log(D) - 20 \log(F)$$

where $D =$ the distance to the target (km) and $F =$ frequency (MHz).

$$P_2 = P_1 - 39 + 10 \log(\sigma) + 20 \log(F)$$

where $\sigma =$ the target's radar cross-section (m²).

$$P_A = P_2 - 32 - 20 \log(D) - 20 \log(F)$$

$$P_R = P_A + G$$

so:

$$P_R = P_T + 2G - 103 - 40 \log(D) - 20 \log(F) + 10 \log(\sigma)$$

2.3.4 Interfering Signals

If two signals at the same frequency arrive at a single antenna, one is usually considered the desired signal and the second is an interfering signal. (See Figure 2.7.) The same equations apply whether the interfering signal is unintentional or intentional jamming. The dB expression for the difference in power between the two signals, assuming that the receiver antenna presents the same gain to both signals, is:

$$P_S - P_I = \mathrm{ERP}_S - \mathrm{ERP}_I - 20\log(D_S) + 20\log(D_I)$$

where P_S is the received power (i.e., at the receiver input) from the desired signal; P_I is the received power from the interfering signal; ERP_S is the effective radiated power of the desired signal; ERP_I is the effective radiated power of the interfering signal; D_S is the path distance to the desired signal transmitter; and D_I is the path distance to the interfering signal transmitter.

This is the simplest form of the interference equation. In Chapter 3, we will deal with directional receiving antennas, which cause different antenna-gain factors to be applied to the two signals. Also, we will, of course, deal with the situation in which an interfering signal (i.e., from a jammer) is accepted by a radar receiver along with the desired radar return signal. All of these expressions will build on the simple dB form expressions described above.

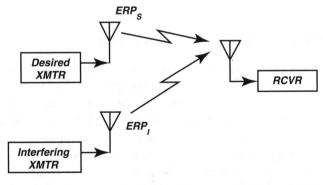

Figure 2.7 Interfering signals can be described in terms of links from each transmitter to the receiver being considered.

2.3.5 Low-Frequency Signals Close to the Earth

The expression for spreading loss given above is typical for EW link applications, but there is another form of this equation that applies to relatively low frequencies transmitted to or from antennas that are close to the Earth.

If the link exceeds the Fresnel zone, the spreading loss obeys the above formula ($L_S = 32 + 20 \log(f) + 20 \log(d)$). The spreading loss for ranges greater than this distance is determined by the formula:

$$L_S = 120 + 40 \log(d) - 20 \log(h_T) - 20 \log(h_R)$$

where L_S = the spreading loss (in dB); d = link distance (in km); h_T = transmitting antenna height (in meters); and h_R = receiving antenna height (in meters).

The distance from the transmitter to the Fresnel zone is calculated from the equation:

$$F_Z = (h_T \times h_R \times f) / 75{,}000$$

where F_Z = the distance to the Fresnel zone (in km); h_T = transmitting antenna height (in meters); h_R = receiving antenna height (in meters); and f = the transmitted frequency (in MHz).

2.4 Relations in Spherical Triangles

Spherical trigonometry is a valuable tool for many aspects of EW, but it will be essential for the consideration of EW modeling and simulation in Chapter 11.

2.4.1 The Role of Spherical Trigonometry in EW

Spherical trig is one way to deal with three-dimensional problems and has the advantage that it deals with spatial relationships from the "point of view" of sensors. For example, a radar antenna typically has an elevation angle and an azimuth angle which define the direction to a target. Another example is the orientation of the boresight of an antenna mounted on an aircraft. With spherical trig, it is practical to define the direction of the boresight in terms of the mounting of the antenna on the aircraft and the orientation of the aircraft in pitch, yaw, and roll. Another example is the determination of

Doppler-shift magnitude when the transmitter and receiver are on two different aircraft with arbitrary velocity vectors.

2.4.2   The Spherical Triangle

A spherical triangle is defined in terms of a unit sphere—that is, a sphere of radius one (1). See Figure 2.8. The origin (center) of this sphere is placed at the center of the Earth in navigation problems, at the center of the antenna in angle-from-boresight problems, and at the center of an aircraft or weapon in engagement scenarios. There are, of course, an infinite number of applications, but for each, the center of the sphere is placed where the resulting trigonometric calculations will yield the desired information.

The "sides" of the spherical triangle must be great circles of the unit sphere—that is, they must be the intersection of the surface of the sphere with a plane passing through the origin of the sphere. The "angles" of the triangle are the angles at which these planes intersect. Both the "sides" and the "angles" of the spherical triangle are measured in degrees. The size of a "side" is the angle the two end points of that side make at the origin of the sphere. In normal terminology, the sides are indicated as lowercase letters, and the angles are indicated with the capital letter corresponding to the side opposite the angle, as shown in Figure 2.9.

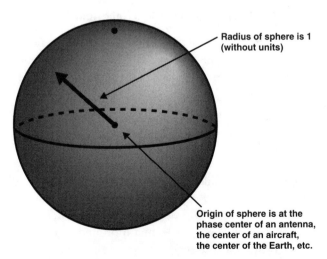

Radius of sphere is 1 (without units)

Origin of sphere is at the phase center of an antenna, the center of an aircraft, the center of the Earth, etc.

Figure 2.8   Spherical trigonometry is based on relationships in a unit sphere. The origin (center) of the sphere is some point that is relevant to the problem being solved.

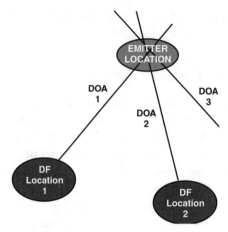

Figure 2.9   A spherical triangle has three "sides" which are great circles of a sphere. It has three "angles" which are the intersection angles of the planes including those great circles.

It is important to realize that some of the qualities of plane triangles do not apply to spherical triangles. For example, all three of the "angles" in a spherical triangle could be 90°.

2.4.3 Trigonometric Relationships in Any Spherical Triangle

While there are many trigonometric formulas, the three most commonly used in EW applications are the Law of Sines, the Law of Cosines for Angles, and the Law of Cosines for Sides. They are defined as follows:

- Law of Sines for Spherical Triangles

$$\sin a/\sin A = \sin b/\sin B = \sin c/\sin C$$

- Law of Cosines for Sides

$$\cos a = \cos b \cos c + \sin b \sin c \cos A$$

- Law of Cosines for Angles

$$\cos A = -\cos B \cos C + \sin B \sin C \cos a$$

Of course, *a* can be any side of the triangle you are considering, and *A* will be the angle opposite that side. You will note that these three formulas are similar to equivalent formulas for plane triangles.

$$a/\sin A = b/\sin B = c/\sin C$$

$$a^2 = b^2 + c^2 - 2bc\cos A$$

$$a = b\cos C + c\cos B$$

2.4.4   The Right Spherical Triangle

As shown in Figure 2.10, a right spherical triangle has one 90° "angle." This figure illustrates the way that the latitude and longitude of a point on the Earth's surface would be represented in a navigation problem, and many EW applications can be analyzed using similar right spherical triangles.

Right spherical triangles allow the use of a set of simplified trigonometric equations generated by Napier's rules. Note that the five-segmented disk in Figure 2.11 includes all of the parts of the right spherical triangle, except the 90° angle. Also note that three of the parts are preceded by "co-." This means that the trigonometric function of that part of the triangle must be changed to the co-function in Napier's rules (i.e., sine becomes cosine, etc.).

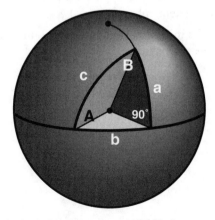

Figure 2.10   A right spherical triangle has one 90° "angle."

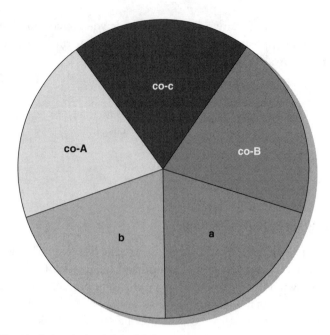

Figure 2.11   Napier's rule for right spherical triangles allow simplified equations in reference to this five-segment circle.

Napier's rules are as follows:

- The sine of the middle part equals the product of the tangents of the adjacent parts. (Remember the co-s.)
- The sine of the middle part equals the product of the cosines of the opposite parts. (Remember the co-s.)

A few example formulas generated by Napier's rules follow.

$$\sin a = \tan b \cotan B$$

$$\cos A = \cotan c \tan b$$

$$\cos c = \cos a \cos b$$

$$\sin a = \sin A \sin c$$

As you will see, when they are applied to practical EW problems, these formulas greatly simplify the math involved with spherical manipulations when you can set up the problem to include a right spherical triangle.

2.5 EW Applications of Spherical Trigonometry

2.5.1 Elevation-Caused Error in Azimuth-Only DF System

A direction-finding (DF) system is designed to measure only the azimuth of arrival of signals. However, signals can be located out of the plane in which the DF sensors assume the emitter is located. What is the error in the azimuth reading as a function of the elevation of the emitter above the horizontal plane?

This example assumes a simple amplitude-comparison DF system. DF systems measure the true angle from the reference direction (typically the center of the antenna baseline) to the direction from which the signal arrives. In an azimuth-only system, this measured angle is reported as the azimuth of arrival (by adding the azimuth of the reference direction to the measured angle).

As shown in Figure 2.12, the measured angle forms a right spherical triangle with the true azimuth and the elevation. The true azimuth is determined as follows:

$$\cos(Az) = \cos(M) \, / \, \cos(El)$$

The error in the azimuth calculation as a function of the actual elevation is then as follows:

$$\text{Error} = M - a\cos[\cos(M) \, / \, \cos(El)]$$

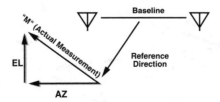

Figure 2.12 The typical DF system measures the angle between the direction from which the signal arrives and a reference direction.

2.5.2 Doppler Shift

Both the transmitter and receiver are moving. Each has a velocity vector with an arbitrary orientation. The Doppler shift is a function of the rate of change of distance between the transmitter and the receiver. To find the rate of change of range between the transmitter and the receiver as a function of the two velocity vectors, it is necessary to determine the angle between each velocity vector and the direct line between the transmitter and the receiver. The rate of change of distance is then the transmitter velocity times the cosine of this angle (at the transmitter) plus the receiver velocity times the cosine of this angle (at the receiver).

Let's place the transmitter and receiver in an orthogonal coordinate system in which the y axis is north, the x axis is west, and the z axis up. The transmitter is located at X_T, Y_T, Z_T; and the receiver is located at X_R, Y_R, Z_R. The directions of the velocity vectors are then the elevation angle (above or below the x,y plane) and azimuth (the angle clockwise from north in the x,y plane), as shown in Figure 2.13. We can find the azimuth and elevation of the receiver (from the transmitter) using plane trigonometry.

$$az_R = A\tan\left[(X_R - X_T)/(Y_R - Y_T)\right] \tag{2.1}$$

$$El_R = a\tan\{(Z_R - Z_T)/SQRT[(X_R - X_T)^2 + (Y_R - Y_T)^2]\} \tag{2.2}$$

Now consider the angular conversions at the transmitter, as shown in Figure 2.14. This is a set of spherical triangles on a sphere with its origin at the transmitter. N is the direction to north; V is the direction of the velocity vector; and R is the direction toward the receiver. The angle from north to the velocity vector can be determined using the right spherical triangle formed by the velocity-vector azimuth and elevation angles. Likewise, the

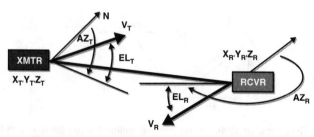

Figure 2.13   For calculation of Doppler shift in the general case, both the transmitter and receiver can have motion with velocity vectors arbitrarily oriented.

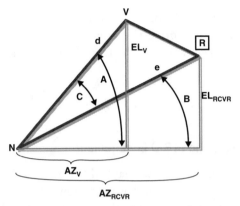

Figure 2.14 On a unit sphere with its origin at the transmitter, there are two right spherical triangles formed by the azimuth and elevation of the velocity vector, and the receiver (as seen from the transmitter).

angle from north to the receiver can be determined from the right spherical triangle formed by its azimuth and elevation:

$$\cos(d) = \cos(Az_V)\cos(El_V)$$

$$\cos(e) = \cos(Az_{RCVR})\cos(El_R)$$

Az_{RCVR} and El_{RCVR} are determined by the method shown in Section 2.5.3. Angles A and B can be determined from:

$$\operatorname{ctn}(A) = \sin(Az_V)/\tan(El_V)$$

$$\operatorname{ctn}(B) = \sin(Az_{RCVR})/\tan(El_R)$$

$$C = A - B$$

Then, from the spherical triangle between N, V, and R, using the law of cosines for sides, we find the angle between the transmitter's velocity vector and the receiver:

$$\cos(VR) = \cos(d)\cos(e) + \sin(d)\sin(e)\cos(C)$$

Now, the component of the transmitter's velocity vector in the direction of the receiver is found by multiplying the velocity by $\cos(VR)$. This same operation is performed from the receiver to determine the component of the receiver's velocity in the direction of the transmitter. The two velocity vectors are added to determine the rate of change of distance between the transmitter and receiver (V_{REL}). The Doppler shift is then found from the following:

$$\Delta f = f V_{REL}/c$$

2.5.3 Observation Angle in 3-D Engagement

Given two objects in three-dimensional (3-D) space, T is a target, and A is a maneuvering aircraft. The pilot of A is facing toward the roll axis of the aircraft, sitting perpendicular to the yaw plane. What are the observed horizontal and vertical angles of T from the pilot's point of view? This is the problem that must be solved to determine where a threat symbol would be placed on a head-up display (HUD).

Figure 2.15 shows the target and the aircraft in the 3-D gaming area. The target is at X_T, Y_T, Z_T; and the aircraft is at X_A, Y_A, Z_A. The roll axis is defined by its azimuth and elevation relative to the gaming-area coordinate system. The azimuth and elevation of the target from the aircraft location are determined as in (2.1) and (2.2) by the following:

$$Az_T = a\tan\left[(X_T - X_A)/(Y_T - Y_A)\right]$$

$$El_T = a\tan\left\{(Z_T - Z_A)/SQRT\left[(X_T - X_A)^2 + (Y_T - Y_A)^2\right]\right\}$$

Note that you need to account for the discontinuities as the angle changes quadrants.

The two right spherical triangles and one spherical triangle of Figure 2.16 allow the calculation of the angular distance from the roll axis and the target (j):

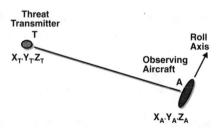

Figure 2.15   A threat emitter is observed by the ESM system in an aircraft.

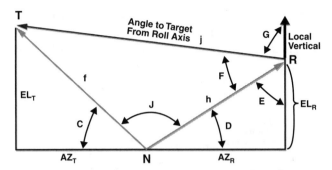

Figure 2.16 On a unit sphere with its origin at the aircraft, there are two right spherical triangles formed by the azimuth and elevation of the target (T) and of the roll axis (R).

$$\cos(f) = \cos(Az_T)\cos(El_T)$$

$$\cos(h) = \cos(Az_R)\cos(El_R)$$

$$\text{ctn}(C) = \sin(Az_T)/\tan(El_T)$$

$$\text{ctn}(D) = \sin(Az_R)/\tan(El_R)$$

$$J = 180° - C - D$$

$$\cos(j)\ 5\ \cos(f)\ \cos(h) + \sin(f)\sin(h)\cos(J)$$

The angle E is then determined:

$$\text{ctn}(E) = \sin(El_R)/\tan(Az_R)$$

The angle F is determined from the law of sines:

$$\sin(F) = \sin(J)\sin f/\sin(j)$$

Then the offset angle of the threat from the local vertical at the aircraft is given by the following:

$$G = 180° - E - F$$

Finally, as shown in Figure 2.17, the location of the threat symbol on the HUD is a distance from the center of the display representing the angular distance (j), and an offset from vertical on the HUD by the sum of angle G and the roll angle of the aircraft from vertical.

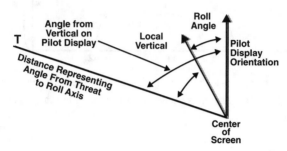

Figure 2.17   The location of the threat display symbol on the operator's screen is determined by the angular distance from the roll axis, and the sum of the angular offset from the threat location to the local vertical and the angular offset caused by the aircraft's roll.

3

Antennas

The purpose of this chapter is not to make you an expert in the antenna field. Rather, it is to provide a general understanding of antennas and the roles and capabilities of various types of antennas. Another purpose is to make you aware of the antenna parameter tradeoffs. After this discussion, you should be able to specify and select antennas and hold reasonably intelligent discussions with the professionals who spend their careers in this highly specialized area.

3.1 Antenna Parameters and Definitions

Antennas impact electronic warfare systems and applications in many ways. In receiving systems, they provide gain and directivity. In many types of direction-finding systems, the antenna parameters are the source of the data from which direction of arrival is determined. In jamming systems, they provide gain and directivity. In threat emitters, particularly radars, the gain pattern and scan characteristics of the transmitting antenna provide one of the important ways to identify the threat signal. The threat emitter antenna scan and polarization also allow the use of some deceptive countermeasures.

This chapter will cover the parameters and common applications for various types of antennas, provide a guide for matching the type of antenna to the job it must do, and offer some simple formulas for the tradeoff of various antenna parameters.

3.1.1 First, Some Definitions

An antenna is any device which converts electronic signals (i.e., signals in cables) to electromagnetic waves (i.e., signals out in the "ether waves")—or vice versa. They come in a huge range of sizes and designs, depending on the frequency of the signals they handle and their operating parameters. Functionally, any antenna can either transmit or receive signals. However, antennas designed for high-power transmission must be capable of handling large amounts of power. Common antenna performance parameters are shown in Table 3.1.

3.1.2 The Antenna Beam

One of the most important (and misstated) areas in the whole EW field has to do with the various parameters defining an antenna beam. Several antenna beam definitions can be described from Figure 3.1, which is the amplitude pattern (in one plane) of an antenna. This can be either the horizontal pat-

Table 3.1
Commonly Used Antenna Performance Parameters

Term	Definition
Gain	The increase in signal strength (commonly stated in dB) as the signal is processed by the antenna. (Note that the gain can be either positive or negative and that an isotropic antenna has unity gain, which is also stated as 0-dB gain.)
Frequency coverage	The frequency range over which the antenna can transmit or receive signals and provide the appropriate parametric performance.
Bandwidth	The frequency range of the antenna in units of frequency. This is often stated in terms of the percentage bandwidth [100% × (maximum frequency − minimum frequency) / average frequency].
Polarization	The orientation of the E and H waves transmitted or received. Mainly vertical, horizontal, or right- or left-hand circular—can also be slant linear (any angle) or elliptical.
Beamwidth	The angular coverage of the antenna, usually in degrees (defined below).
Efficiency	The percentage of signal power transmitted or received compared to the theoretical power from the proportion of a sphere covered by the antenna's beam.

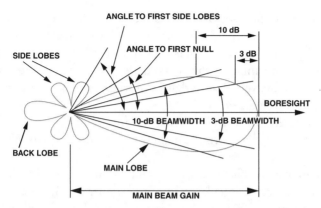

Figure 3.1 Antenna parameter definitions are based on the geometry of the antenna gain pattern.

tern or the vertical pattern. It can also be the pattern in any other plane which includes the antenna. This type of pattern is made in an anechoic chamber designed to prevent signals from reflecting off its walls. The subject antenna is rotated in one plane while receiving signals from a fixed test antenna, and the received power is recorded as a function of the antenna's orientation relative to the test antenna.

Boresight: The boresight is the direction the antenna is designed to point. This is usually the direction of maximum gain, and the other angular parameters are typically defined relative to the boresight.

Main lobe: The primary or maximum gain beam of the antenna. The shape of this beam is defined in terms of its gain versus angle from boresight.

Beamwidth: This is the width of the beam (usually in degrees). It is defined in terms of the angle from boresight that the gain is reduced by some amount. If no other information is given, "beamwidth" usually refers to the 3-dB beamwidth.

3-dB beamwidth: The two-sided angle (in one plane) between the angles at which the antenna gain is reduced to half of the gain at the boresight (i.e., 3-dB gain reduction). Note that all beamwidths are "two-sided" values. For example, in an antenna with a 3-dB beamwidth of 10° the gain is 3 dB down 5° from the boresight, so the two 3-dB points are 10° apart.

"n" dB beamwidth: The beamwidth can be defined for any level of gain reduction. The 10-dB beamwidth is shown in the figure.

Side lobes: Antennas have other than intended beams as shown in the figure. The back lobe is in the opposite direction from the main beam, and the side lobes are at other angles.

Angle to the first side lobe: This is the angle from the boresight of the main beam to the maximum gain direction of the first side lobe. Note that this is a single-sided value. (It makes people crazy the first time they see a table in which the angle to the first side lobe is less than the main beam beamwidth—before they realize that beamwidth is two sided and the angle to the side lobe is single sided.)

Angle to the first null: This is the angle from the boresight to the minimum-gain point between the main beam and the first side lobe. It is also a single-sided value.

Side-lobe gain: This is usually given in terms of the gain relative to the main-beam boresight gain (a large negative number of dB). Antennas are not designed for some specific side-lobe level—the side lobes are considered bad, and thus certified by the manufacturer to be below some specified level. However, from an EW or reconnaissance point of view, it is important to know the side-lobe level of the transmitting antennas for signals you want to intercept. EW receiving systems are often designed to receive "0-dB side lobes"—which is to say that the side lobes are down from the main lobe gain by the amount of that gain. For example, "0 dB" side lobes from a 40-dB gain antenna would transmit with 40 dB less power than observed if the antenna boresight is pointed directly at your receiving antenna.

3.1.3   More About Antenna Gain

In order to just add the antenna gain to a received signal's strength, we need to state signal strength out in the "ether waves" in dBm—which is not really true. As discussed in Chapter 2, dBm is really a logarithmic representation of power in milliwatts—which only occurs in a circuit. The strength of a transmitted signal is more accurately stated in microvolts per meter (μv/m) of field strength, and the sensitivity of receivers with integral antennas is often stated in μv/m. That same column gives convenient formulas for the conversion between dBm and μv/m.

3.1.4   About Polarization

From an EW point of view, the most important effect of polarization is that the power received in an antenna is reduced if it does not match the polarization of the received signal. In general (but not always), linearly polarized antennas have geometry which is linear in the polarization orientation (e.g., vertically polarized antennas tend to be vertical). Circularly polarized antennas tend to be round or crossed, and they can be either right-hand or left-hand circular (LHC or RHC). The gain reduction from various polarization matches is shown in Figure 3.2.

An important EW polarization trick is to use a circularly polarized antenna to receive a linearly polarized signal of unknown orientation. You always lose 3 dB but avoid the 25-dB loss that would occur if you were cross polarized. When the received signal can have any polarization (i.e., any linear or either circular), it is common practice to make quick measurements with LHC and RHC antennas and choose the stronger signal. The value of 25 dB for cross-polarized antennas is normal for the types of antennas common to EW systems (usually covering wide frequency ranges). Narrow-frequency-band antennas (for example, in communication satellite links) can be carefully designed for cross-polarization isolation of greater than 30 dB. The small, circularly polarized antennas in RADAR warning receiver systems can have as little as 10 dB cross-polarization isolation.

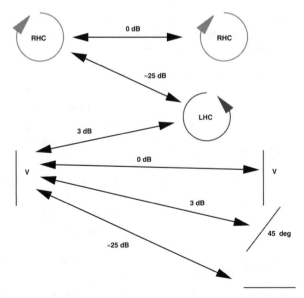

Figure 3.2 Cross-polarization losses range from 0 dB to approximately 25 dB. Note that the 3-dB loss applies between any linear and circular polarization combination.

3.2 Types of Antennas

There are many types of antennas used in EW applications. They vary in their angular coverage, in the amount of gain they provide, in their polarization, and in their physical size and shape characteristics. Selection of the best type of antenna is highly application-dependent, often requiring difficult performance tradeoffs with significant impact on other system design parameters.

3.2.1 Selecting an Antenna to Do the Job

To do what is required for any specific EW application, an antenna must provide the required angular coverage, polarization, and frequency bandwidth. Table 3.2 provides an antenna selection guide in terms of those general performance parameters. In this table, the angular coverage is just divided between "360° azimuth" and "directional." Antennas with 360° azimuth

Table 3.2

Selection of Antenna Type for Either Transmission or Reception Is Made Based on Its Angular Coverage, Polarization, and Bandwidth

Angular Coverage	Polarization	Bandwidth	Antenna Type
360° azimuth	Linear	Narrow	Whip, dipole, or loop
		Wide	Biconical or swastika
	Circular	Narrow	Normal mode helix
		Wide	Lindenblad or four-arm conical spiral
Directional	Linear	Narrow	Yagi, array with dipole elements, or dish with horn feed
		Wide	Log periodic, horn, or dish with log periodic feed
	Circular	Narrow	Axial mode helix or horn with polarizer or dish with crossed dipole feed
		Wide	Cavity-backed spiral, conical spiral, or dish with spiral feed

coverage are often called "omnidirectional," which is not true. An omni-directional antenna would provide consistent spherical coverage, whereas these types of antennas provide only a limited elevation coverage (some more limited than others). Still, they are "omni" enough for most applications in which signals from "any direction" must be received at any instant—or for which it is desirable (or acceptable) that signals be transmitted in all directions. Directional antennas provide limited coverage in both azimuth and elevation. Although they must be pointed toward the desired transmitter or receiver location, they typically provide more gain than the 360° types. Another advantage of directional antennas is that they significantly reduce the level at which undesired signals are received—or alternately the effective radiated power transmitted to hostile receivers.

Next, the table differentiates by polarization, and finally it differentiates by frequency bandwidth (only narrow or wide). In most EW applications, "wide" bandwidth means an octave or more (sometimes much more).

3.2.2 General Characteristics of Various Types of Antennas

Figure 3.3 is a convenient summary of parameters of the various types of antennas used in EW applications. For each antenna type, the left-hand column shows a rough sketch of the antenna's physical characteristics. The center column shows a very general elevation and azimuth gain pattern for that antenna type. Only the general shape of these curves is useful—the actual gain pattern for a specific antenna of that type will be determined by its design. The right-hand column is a summary of the *typical* specifications to be expected. Typical is an important word here, since the possible range of parameters is much wider. For example, it is theoretically possible to use any antenna type in any frequency range. However, practical considerations of physical size, mounting, and appropriate applications cause a particular antenna type to be used in that "typical" frequency range.

Antenna Type	Pattern	Typical Specifications
Dipole	El / Az	Polarization: Vertical Beamwidth: 80° x 360° Gain: 2 dB Bandwidth: 10% Frequency Range: zero through μw
Whip	El / Az	Polarization: Vertical Beamwidth: 45° x 360° Gain: 0 dB Bandwidth: 10% Frequency Range: HF through UHF
Loop	El / Az	Polarization: Horizontal Beamwidth: 80° x 360° Gain: -2 dB Bandwidth: 10% Frequency Range: HF through UHF
Normal Mode Helix	El / Az	Polarization: Horizontal Beamwidth: 45° x 360° Gain: 0 dB Bandwidth: 10% Frequency Range: HF through UHF
Axial Mode Helix	Az & El	Polarization: Circular Beamwidth: 50° x 50° Gain: 10 dB Bandwidth: 70% Frequency Range: UHF through lowμw
Biconical	El / Az	Polarization: Vertical Beamwidth: 20° x 100° x 360° Gain: 0 to 4 dB Bandwidth: 4 to 1 Frequency Range: UHF through mmw
Lindenblad	El / Az	Polarization: Circular Beamwidth: 80° x 360° Gain: -1 dB Bandwidth: 2 to 1 Frequency Range: UHF through μw
Swastika	El / Az	Polarization: Horizontal Beamwidth: 80° x 360° Gain: -1 dB Bandwidth: 2 to 1 Frequency Range: UHF through μw
Yagi	El / Az	Polarization: Horizontal Beamwidth: 90° x 50° Gain: 5 to 15 dB Bandwidth: 5% Frequency Range: VHF through UHF
Log Periodic	El / Az	Polarization: Vertical or Horizontal Beamwidth: 80° x 60° Gain: 6 to 8 dB Bandwidth: 10 to 1 Frequency Range: HF through μw

Antenna Type	Pattern	Typical Specifications
Cavity Backed Spiral	Az & El	Polarization: R & L Horizontal Beamwidth: 60° x 60° Gain: -15 dB (min freq) +3 dB (max freq) Bandwidth: 9 to 1 Frequency Range: μw
Conical Spiral	Az & El	Polarization: Circular Beamwidth: 60° x 60° Gain: 5 to 8 dB Bandwidth: 4 to 1 Frequency Range: UHF through μw
4 Arm Conical Spiral	El / Az	Polarization: Circular Beamwidth: 50° x 360° Gain: 0 dB Bandwidth: 4 to 1 Frequency Range: UHF through μw
Horn	El / Az	Polarization: Linear Beamwidth: 40° x 40° Gain: 5 to 10 dB Bandwidth: 4 to 1 Frequency Range: VHF through mmw
Horn with Polarizer	El / Az	Polarization: Circular Beamwidth: 40° x 40° Gain: 4 to 10 dB Bandwidth: 3 to 1 Frequency Range: μw
Parabolic Dish	Az & El	Polarization: Depends on Feed Beamwidth: 0.5° x 30° Gain: 10 to 55 dB Bandwidth: Depends on Feed Frequency Range: UHF to μw
Phased Array	El / Az	Polarization: Depends on Elements Beamwidth: 0.5° x 30° Gain: 10 to 40 dB Bandwidth: Depends on Elements Frequency Range: VHF to μw
	Elements	

Figure 3.3 Each type of antenna has a characteristic gain pattern and typical specifications; the patterns and specifications of specific antennas of each type are determined by their design detail.

3.3 Parameter Tradeoffs in Parabolic Antennas

One of the most flexible types of antennas used in EW applications (and many others) is the parabolic dish. The definition of a parabolic curve is such that it reflects rays from a single point (the focus) to parallel lines. By placing a transmitting antenna (called a feed) at the focus of a parabolic dish, we can direct all of the signal power which hits the dish in the same direction (theoretically). An ideal feed antenna will radiate all of its energy to the dish. (Actually, transmitting about 90% of the energy onto the dish is considered ideal enough for most practical purposes.) The actual antenna pattern will generate a main lobe, which rolls off in an angle, a back lobe, and side lobes.

There is a relationship between the size of an antenna's reflector, the operating frequency, the efficiency, the effective antenna area, and the gain. This relationship is presented below in several useful forms.

3.3.1 Gain Versus Beamwidth

Figure 3.4 shows the gain versus the beamwidth of a parabolic antenna with 55% efficiency. This efficiency is what you expect in a commercially available antenna which covers a relatively small frequency bandwidth (about 10%).

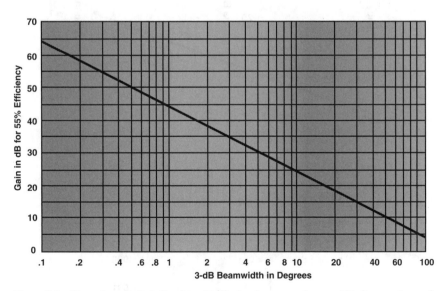

Figure 3.4   There is a well-defined tradeoff of gain versus beamwidth for any type of antenna. This chart shows the gain versus beamwidth for a parabolic antenna with 55% efficiency.

For the wide bandwidth antennas often used in EW and reconnaissance applications (an octave or more), the efficiency will be less than 55%. The beam is assumed to be symmetrical in azimuth and elevation. To use the table, draw a line from the antenna beamwidth up to the line and then left to the gain in dB.

3.3.2 Effective Antenna Area

Figure 3.5 is a nomograph of the relationship between the operating frequency, the antenna boresight gain, and the effective antenna area. The line

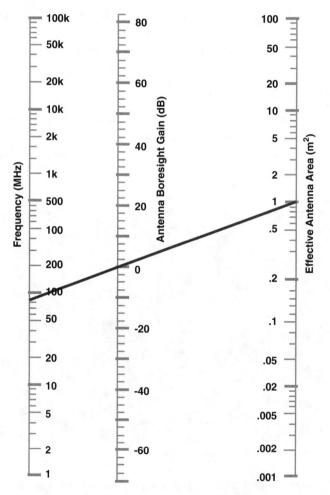

Figure 3.5   The effective area of an antenna is a function of its gain and the operating frequency.

on the figure is for an isotropic antenna (0-dB gain) with a $1\,m^2$ effective area. You can see that this occurs at approximately 85 MHz. The equation for this nomograph is:

$$A = 38.6 + G - 20 \log (F)$$

where A is the area in dBsm (i.e., dB relative to $1m^2$); G is the boresight gain (in dB); and F is the operating frequency (in MHz).

3.3.3 Antenna Gain as a Function of Diameter and Frequency

Figure 3.6 is a nomograph which can be used to determine the gain of an antenna from its diameter and the operating frequency. Note that this is specifically for a 55% efficient antenna. The line on the figure shows that a 0.5-m-diameter antenna with 55% efficiency will have a gain of approximately 32 dB at 10 GHz. This nomograph assumes that the surface of the dish is a parabola accurate to a small part of a wavelength at the operating frequency; otherwise, there is a gain reduction. The equation for this nomograph is:

$$G = -42.2 + 20 \log (D) + 20 \log (F)$$

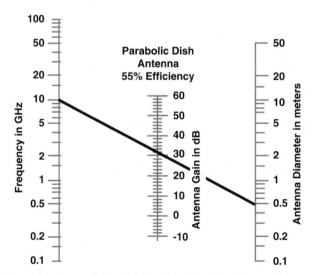

Figure 3.6 The gain of a parabolic dish is a function of its diameter, the operating frequency, and the antenna efficiency. This nomograph is for 55% efficiency.

where G is the antenna gain (in dB); D is the reflector diameter (in meters); F is the operating frequency (in MHz).

Several antenna manufacturers will provide you with handy slide rules that can be used to perform this tradeoff (for any efficiency) if you call or write to their sales offices. (They are typically free advertising handouts.) These slide rules also include other useful information (and are fun to play with).

Table 3.3 shows adjustment of gain as function of antenna efficiency. Since Figures 3.4 and 3.6 assume 55% efficiency, this table is very useful in adjusting the determined gain numbers for other values of efficiency.

3.3.4 Gain of Nonsymmetrical Antennas

The above discussion assumes that the antenna beam is symmetrical (i.e., azimuth and elevation of the beam are equal). The gain of a 55% efficient parabolic dish with a nonsymmetrical pattern can be determined from the equation:

$$\text{Gain (not in dB)} = [29,000/(\theta_1 \times \theta_2)]$$

where θ_1 and θ_2 are the 3-dB beamwidth angles in two orthogonal directions (e.g., vertical and horizontal).

Naturally, this is converted to the gain in dB by taking $10 \times$ the log of the value on the right side of the equation.

This equation is empirical, but it can be derived (pretty closely) by assuming that the gain is equivalent to the energy concentration within the 3-dB beamwidth. Thus, the gain is equivalent to the ratio of the surface of a sphere to the surface area inside an ellipse on the sphere's surface with major and minor axes (dimensioned in spherocentric degrees) equal to the two angles describing the antenna beam coverage (remember the 55% efficiency factor).

Table 3.3
The Gain of a 55% Efficient Antenna Can Be Modified for Different Efficiencies
Using This Table

Antenna Efficiency	Adjustment to Gain (vs. 55%)
60%	Add 0.4 dB
50%	Subtract 0.4 dB
45%	Subtract 0.9 dB
40%	Subtract 1.4 dB
35%	Subtract 2 dB
30%	Subtract 2.6 dB

3.4 Phased Array Antennas

For a number of extremely practical reasons, phased array antennas are becoming increasingly important to the field of electronic warfare (EW). In radars, phased arrays can be instantly switched from one target to another, increasing the efficiency with which multiple targets can be acquired and/or tracked. From the EW point of view, this makes it (in general) impossible to determine the threat radar's antenna parameters from an analysis of the time history of the received signal strength.

When phased arrays are used for receiving or jamming antennas, the EW system gains the same flexibility enjoyed by the threat radar. For example, a jammer can split its jamming power among multiple threats and/or instantly move from one to the other. In some applications, it will be practical to simultaneously receive and jam from the same array.

The phased array will reach its ultimate usefulness to EW when so-called "smart skin" technology is realized on aircraft. This is a scheme in which most or all of the skin of the aircraft contains parts of antenna elements which can be configured into massive phased arrays.

An additional advantage of the phased array is that it can be made conformal to the shape of the vehicle carrying it. Anyone who has dealt with the aerodynamic problems of a mechanically scanned antenna on an aircraft will appreciate the ability of the phased array to conform to the skin of the aircraft. The aerodynamics folks get really ugly when you want to extend your radome to allow a parabolic antenna to be directed over a wide angle of view.

As you may have guessed, these wonderful advantages do not come without paying a price (in performance as well as money). The following is a set of general guidelines as to the performance limitations and design constraints for phased array antennas. For more information, you are directed to any of the texts listed on page 3.

3.4.1 Phased Array Antenna Operation

As shown in Figure 3.7, a phased array is a group of antennas, each of which is connected to a phase shifter. When used as a transmitting antenna, the signal to be transmitted is divided among the antennas, and the phase of the signal to each antenna is adjusted so that all of the signals will be in phase when viewed from some selected direction—and will thus add constructively. It follows that they will be out of phase, when viewed from any other angle and will thus add less constructively. This forms an antenna beam.

When used as a receiving antenna array, the phase shifters cause a signal received from the selected direction to add in phase at the signal combiner.

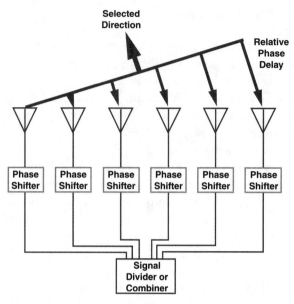

Figure 3.7 A phased array antenna comprises a number of antenna elements, each connected to an individually controlled phase shifter. The phase shifters are set so that a signal arriving from a selected angle will add in phase at the signal combiner—or conversely so that the signals transmitted from all elements will add constructively when viewed from a selected angle.

There are linear arrays in which the antennas are in a single line—which is narrowed and directed by the phase shifters in a plane (for example, the horizontal plane). In this case the beamwidth of the array is determined by the phase shifters only in that plane. The beamwidth in the other orientation (for example, the vertical) is determined by the beamwidth of the individual antennas in that dimension.

There are also planar arrays in which the antennas are arrayed both vertically and horizontally to provide control of both the vertical and horizontal beam width and steering by the phase shifters.

It will be noted that a phase shift will cause a distance delay equal to:

Signal wavelength (phase shift /360°)

For operation over a wide frequency bandwidth, the phase shifters may actually be so-called "true time delay" devices, which delay the signal by a physical distance that is independent of signal frequency.

Like any other type of antenna, the phased array has a beamwidth and a gain that are interactive.

3.4.2 Antenna Element Spacing

In general, the individual antennas making up a phased array should be spaced by one-half wavelength at the highest frequency, as shown in Figure 3.8. This avoids "grating lobes," which degrade the antenna's performance when it is steered.

3.4.3 Phased Array Antenna Beamwidth

The 3-dB beamwidth of a phased array with dipole elements spaced at one-half wavelength is determined by the formula:

$$\text{Beamwidth} = 102/N$$

where N is the number of elements in the array, and the beamwidth is in degrees.

For example, the horizontal beamwidth of a 10-element horizontal linear array would be 10.2°. This is the beamwidth in the direction perpendicular to the orientation of the arrayed antennas. For arrays of higher gain antennas, the beamwidth is the element beamwidth divided by N.

As shown in Figure 3.9, this beamwidth increases by the cosine of the angle as it is steered away from the array's boresight angle. In the case of

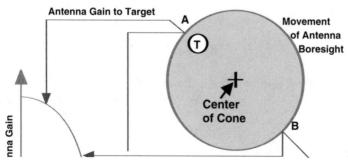

Figure 3.8 To avoid grating lobes, antenna element spacing must not be greater than one-half wavelength at the highest frequency.

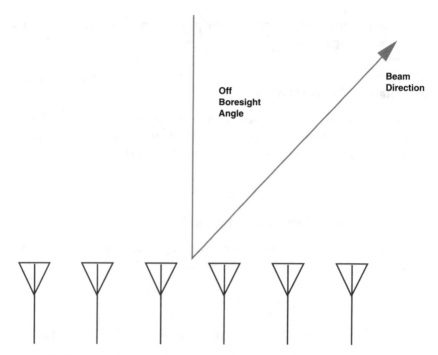

Figure 3.9   The gain of the array is reduced by the cosine of the off-boresight angle. The beamwidth is increased by the same ratio.

our 10.2° beam, it would increase to 14.4° if it were steered 45° from boresight.

3.4.4   Phased Array Antenna Gain

The gain of a phased array with half-wavelength element spacing is given by the formula:

$$G = 10 \log_{10}(N) + Ge$$

where G is the array gain (when it is directed 90° to the line of antennas); N is the number of elements in the array; and Ge is the gain of an individual element.

For example, if the gain of each element were 6 dB and there were 10 elements, the array gain would be 16 dB. Referring again to Figure 3.9, this gain is reduced by the cosine of the angle from boresight—but this is the gain factor, not the gain in dB. In dB, the gain reduction factor will be

$$10 \log_{10} \ (\textit{cosine of angle from boresight})$$

The gain reduction for 45° from boresight is then 0.707 or 1.5 dB.

3.4.5 Beam Steering Limitation

A phased array with half-wavelength element spacing can only be steered approximately 45° from boresight. If the elements are closer (reducing the boresight gain), it can be steered to 60°.

4

Receivers

Receivers are an important part of almost every kind of electronic warfare system. There are many types of receivers, and their characteristics determine their roles. This chapter will first describe the most important types of receivers for EW applications. Then, it will describe receiver systems in which multiple types of receivers are used for a single application. Finally, it will cover the calculation of sensitivity for various types of receivers.

Table 4.1
Types of Receivers Commonly Used in EW Systems

Receiver Type	General Characteristics
Crystal video	Wideband instantaneous coverage; low sensitivity and no selectivity; mainly for pulsed signals
IFM	Coverage, sensitivity, and selectivity like crystal video; measures frequency of received signals
TRF	Like crystal video, but provides frequency isolation and slightly better sensitivity
Superheterodyne	Most common type of receiver; good selectivity and sensitivity
Fixed tuned	Good selectivity and sensitivity; dedicated to one signal
Channelized	Combines selectivity and sensitivity with wideband coverage
Bragg cell	Wideband instantaneous coverage; low dynamic range; multiple simultaneous signals; does not demodulate
Compressive	Provides frequency isolation; measures frequency; does not demodulate
Digital	Highly flexible; can deal with signals with unknown parameters

The ideal EW receiver would be able to detect all types of signals at all frequencies with very good sensitivity 100% of the time. It would be able to detect and demodulate multiple simultaneous signals, including very weak signals in the presence of very strong signals. It would also be small, light, and inexpensive and draw very little power.

Unfortunately, such a receiver has yet to be developed. Most complex systems combine several receiver types to achieve optimum results for specific types of anticipated signal environments. Table 4.1 lists the nine most common types of receivers used in EW systems, along with the general characteristics of each. Table 4.2 lists the specific capabilities of each type.

In general, crystal video and instantaneous frequency measurement (IFM) receivers are used for low- to medium-cost systems operating in high-density pulse signal environments. Both provide 100% coverage of wide

Table 4.2
Characteristics of EW Receivers

Receiver type	Receives pulse	Receives CW	Measures frequency	Selectivity	Multiple signals	Sensitivity	Frequency coverage	Probability of intercept	Dynamic range	Demodulates signals
Crystal video	Y	N	N	P	N	P	G	G	G	Y
IFM	Y	Y	Y	P	N	P	G	G	M	N
TRF	Y	Y	Y	M	Y	P	G	P	G	Y
Superheterodyne	Y	Y	Y	G	Y	G	G	P	G	Y
Fixed tuned	Y	Y	Y	G	Y	G	P	P	G	Y
Channelized	Y	Y	Y	G	Y	G	G	G	G	Y
Bragg cell	Y	Y	Y	G	Y	M	G	G	P	N
Compressive	Y	Y	Y	G	Y	G	G	G	G	N
Digital	Y	Y	Y	G	Y	G	G	M	G	Y

G = Good   M = Moderate   P = Poor   Y = Yes   N = No

frequency ranges, but can't handle multiple simultaneous signals with any dignity. Thus, a high-power CW signal anywhere in their frequency ranges seriously degrades their ability to receive pulses. Also, they have low sensitivity, so they work best against very strong signals. In modern systems, they are often combined with narrowband types of receivers to handle problem situations.

Since fixed-tuned and superheterodyne receivers are narrowband, they are often combined with other types of receivers to isolate simultaneous signals and improve sensitivity. Tuned radio frequency (TRF) receivers also isolate simultaneous signals. The problem with these types is, of course, that they cover only narrow parts of the frequency spectrum at any one instant, creating a low probability of receiving unanticipated signals.

Bragg cell and compressive receivers provide instantaneous coverage of wide frequency ranges and can handle multiple simultaneous signals, but they do not demodulate the signals.

Channelized and digital receivers are the wave of the future. They provide most of the receiver performance parameters that EW systems require, but their size, weight, and power specs reflect the state of the art in component and subsystem miniaturization. At the current state of the art, both types require so much size, weight, and power and are so expensive that they perform only the hardest part of the job in fairly complex systems.

Now, the specific receiver types. . . .

4.1 Crystal Video Receiver

A crystal video receiver is the simplest type of receiver in use today. It consists of a crystal (diode) detector followed by a video amplifier. It amplitude demodulates every signal input to the detector, from DC (unless the detector is AC-coupled to the amplifier) to very high microwave frequencies. The amplitude modulation from all of these signals is combined in the video amplifier and output.

To be useful, the crystal video receiver typically follows a bandpass filter so that only the signals in some band of interest (for example, 2–4 GHz) are received and output. Typically, a log video amplifier is used in this receiver type to give it wide dynamic range.

The signals input to the crystal detector are low enough in power that the detector operates in the "square law" region—that is, the output is a function of the input power rather than the signal voltage. (In other types of receivers, the detection happens at about 10 mW, allowing the detector to operate in the "linear" region.) In his classic 1956 paper on crystal video receivers, Dr. Bill Ayer includes a chart that shows sensitivity for 0-dB signal-

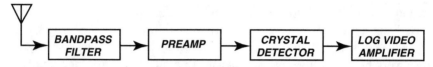

Figure 4.1   Crystal video receivers are normally used with bandpass filters and preamplifiers to tailor their frequency coverage and improve their frequency coverage and sensitivity.

to-noise (SNR) ratio $\approx -54 + 5\log_{10} B_V$ dBm for "good" 1956-vintage detector diodes (where B_V is the video bandwidth in megahertz). Considering that most EW systems depend on automatic pulse processing (which requires 15 or more dB SNR) and must have bandwidth wide enough to handle the narrowest anticipated pulses, a good rule of thumb today is -40 to -45 dBm sensitivity for a crystal video receiver.

The output of the crystal video receiver is a series of pulses with amplitude proportional to the received signal power of each received RF pulse, and with the same start and stop times. When two received pulses overlap, the output will be the combination of both. A strong in-band CW signal will combine with all pulses to distort their amplitudes in the video output.

As shown in Figure 4.1, crystal video receivers normally follow both bandpass filters and preamplifiers. With optimum preamplifier gain (also defined in Dr. Ayer's paper), the sensitivity of the preamplified crystal video receiver is:

$$S_{max} = -114 \text{ dBm} + N_{PA} + 10\log_{10}(B_e) + \text{SNR}_{RQD}$$

where S_{max} = the sensitivity (in dBm) with optimum preamplifier gain; N_{PA} = the preamplifier noise figure (in dB); B_e = the effective bandwidth (in MHz) = $(2B_r B_v - B_{v2})^{1/2}$; and SNR_{RQD} = the required signal-to-noise ratio (in dB).

For a modern crystal video receiver in a typical configuration, with an automatically processed output, the ultimate sensitivity is improved to the -65 to -70 range by preamplification.

4.2   IFM Receiver

The instantaneous frequency measurement (IFM) receiver does just what its name implies. The basic IFM circuit produces a pair of signals that are a function of the radio frequency of the received signal. These signals are digitized to produce a direct digital frequency reading. As shown in Figure 4.2, the input is band limited. A delay line in the IFM circuit sets its output range

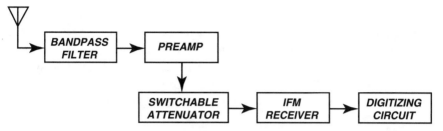

Figure 4.2 Instantaneous frequency measurement receivers provide digital readings of the RF frequency of pulse or CW signals.

to unambiguously cover the input frequency band with maximum accuracy. Since the IFM circuit is also sensitive to signal level, the input to the IFM receiver is first passed through a hard limiting amplifier to produce a constant signal level.

A preamplified IFM receiver has approximately the same sensitivity but somewhat less dynamic range than a crystal video receiver. The switchable attenuator in Figure 4.2 extends its dynamic range to the equivalent of that of the crystal video receiver. An IFM receiver typically measures signal frequency to approximately 1/1,000 of its input frequency range (e.g., 2-MHz resolution over 2–4 GHz). It is fast enough to measure frequency during very short pulses (small part of a microsecond), but gives meaningless readings if more than one signal of comparable strength is present. A strong in-band CW signal precludes the IFM from accurately measuring the frequency of any pulses.

4.3 Tuned Radio Frequency Receiver

Early in the development of radio, many receivers used tuned radio frequency (TRF) designs. They had multiple stages of tuned filtering and gain at the actual frequency of the signal being received. The simplicity of the superheterodyne approach has largely replaced true TRF receiver designs. However, there is another approach, as shown in Figure 4.3, used in EW receiver design that is sometimes called TRF.

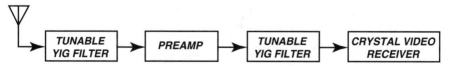

Figure 4.3 Crystal video receivers used with tunable YIG filters to isolate simultaneous signals are often called tuned radio frequency receivers.

The TRF receiver is basically a crystal video receiver with its input frequency range limited by a tuned YIG bandpass filter. This allows the crystal video receiver to handle multiple simultaneous signals and also gives it somewhat better sensitivity because of the narrowed RF bandwidth. In system applications, the TRF receiver may be preceded by an additional preamplifier and a switched attenuator to extend its dynamic range.

4.4 Superheterodyne Receiver

The superheterodyne receiver is extremely flexible. Because it uses a linear detector or discriminator, it provides the best available sensitivity as a function of its predetection bandwidth and postdetection processing gain. The basic superheterodyne receiver "heterodynes" (i.e., linearly shifts) a portion of its RF frequency range into a fixed intermediate frequency (IF) band using a tuned local oscillator (LO). A fixed IF is much more efficient for the provision of the necessary gain and filter selectivity.

Isolation from interfering signals is achieved by the addition of a tuned "preselector" filter, which is controlled along with the local oscillator to select only the portion of the input spectrum that is converted to the IF bandwidth. A simple superheterodyne receiver with tuned preselection is shown in Figure 4.4.

By adjusting the preselector and IF bandwidths, an optimum combination of sensitivity, selectivity, and instantaneous frequency spectrum coverage is achievable. More complex superheterodyne receiver designs, including multiple conversions, are sometimes required to cover large frequency ranges or provide large amounts of isolation in challenging signal environments. Receivers often have selectable IF bandwidths and selectable detectors/discriminators to handle different signal modulations.

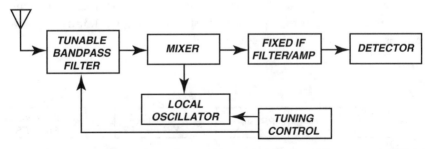

Figure 4.4 Superheterodyne receivers provide optimum tradeoff of sensitivity, selectivity, and bandwidth depending on the selection of filter parameters.

Superheterodyne receivers are normally chosen for EW and reconnaissance systems that are basically narrowband in nature (for example, communications-band ESM systems and many ELINT collection systems). They are also added to wideband systems to handle difficult situations (e.g., detailed parametric analysis of CW signals).

4.5 Fixed Tuned Receiver

In any case in which a single signal (or multiple signals that are always at a single frequency) must be monitored, a fixed tuned receiver may be appropriate. This is typically a true TRF receiver or a superheterodyne receiver with a preset LO. In either case, the simple receiver provides 100% probability of intercept at a single frequency.

4.6 Channelized Receiver

A set of fixed frequency receivers with their passbands set contiguously (usually with the upper edge of the 3-dB bandwidth of one receiver at the same frequency as the lower edge of the 3-dB bandwidth of the next) is called a channelized receiver (Figure 4.5). This is one of the ideal receiver types. It provides a demodulated output for signals in each channel. It can have narrow bandwidth to provide excellent sensitivity and selectivity. It provides 100% probability of intercept for signals within its frequency range, and it can, of course, provide full feature reception for multiple simultaneous signals as long as they are in different frequency channels.

The problem, of course, is in the complexity of implementation. If you want 1 MHz of isolation across the frequency range 2–4 GHz, you will need to have 2,000 channels. That's 2,000 separate receivers—requiring 2,000

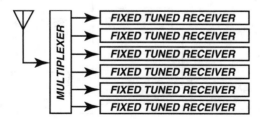

Figure 4.5   A channelized receiver is a set of fixed tuned receivers covering a frequency range to provide 100% receipt and detection of multiple simultaneous signals.

times the size, weight, and power of a single receiver. The good news is that packaging technology is moving in the right direction. Miniaturization technology is bringing the size, weight, power, and cost per channel down at an impressive rate—but these values have not reached the point at which channelized receivers can be used with impunity.

A typical channelized receiver has 10 or 20 channels covering 10% or 20% of the frequency range that an EW system must handle. By use of a switchable frequency translator, slices of the system's frequency range are selected and shifted to the frequency band covered by a single channelized receiver. The channelized receiver is thus applied to solve hard problems (e.g., CW signals, multiple simultaneous signals, or particularly critical parameters) wherever they occur in the EW system's frequency range. It is a valuable asset that is carefully used (under computer control) according to a well-established priority scheme.

4.7 Bragg Cell Receiver

The Bragg cell receiver shown in Figure 4.6 is an instantaneous spectrum analyzer capable of handling multiple simultaneous signals. RF signals amplified to a high power level are applied to a crystal "Bragg cell," which reacts by generating internal compression lines spaced proportionally to the wavelength of any RF signal present in the receiver input. This causes a laser beam to be defracted at an angle proportional to any RF frequency present. This set of defracted beams is focused on a light-detecting array. The array detects the deflection angles of all components of the diffracted beam and produces

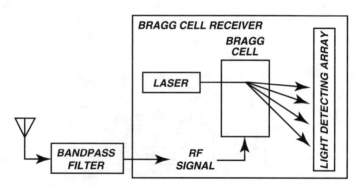

Figure 4.6   Bragg cell receivers provide instantaneous, full-band frequency measurement and handle multiple, simultaneous signals.

output signals from which a digital readout of all signal frequencies present in the input can be determined.

In application, the Bragg cell receiver is used to determine the frequency of signals present so that narrowband receivers can be rapidly tuned to handle them. Its sensitivity is of the same order of magnitude as superheterodyne receivers with the same frequency resolution.

The Bragg cell receiver has limited dynamic range—a problem that has been "just about to be solved" for more than 30 years now. Although it is appropriate for some applications, the Bragg cell technique is being overtaken by the steady march of the state of the art in channelized and digital receivers.

4.8 Compressive Receiver

Figure 4.7 shows the block diagram of a compressive receiver—also called a microscan receiver. It is basically a superheterodyne receiver that is rapidly tuned. Normally, a superheterodyne receiver (or any other type of narrowband receiver) is only tunable at a rate that allows its bandwidth to dwell at a single frequency for a period equal to or greater than its bandwidth (i.e., a receiver with 1-MHz bandwidth must dwell at each frequency for at least 1 µsec). The compressive receiver's tuning rate is much faster than that rate, but its output is passed through a compressive filter that has a delay proportional to frequency. The delay versus frequency slope exactly compensates for the receiver's sweep rate. Thus, as the receiver sweeps its bandwidth across a signal, the output of the receiver is coherently time compressed to make a strong signal spike. The resultant output is a spectral display of the full band over which the receiver tunes.

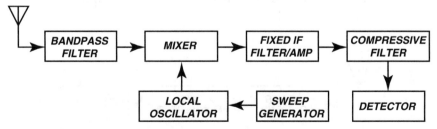

Figure 4.7 A compressive receiver sweeps much faster than the normal one-bandwidth limit, and uses a matched compressive filter to integrate received signals to measure the frequency of all signals within the receiver's frequency range.

Like the Bragg cell, the compressive receiver provides 100% probability of intercept for multiple simultaneous signals and has sensitivity equivalent to that of a regular superheterodyne receiver with the same frequency resolution, but it provides better dynamic range. Also like the Bragg cell, it cannot demodulate signals, and is thus typically most useful in detecting new signals for handoff to narrowband receivers.

4.9 Digital Receivers

The digital receiver seems to be the great hope of the future (Figure 4.8). Basically, it *digitizes* signals for processing in a computer. Since software can functionally simulate any type of filter or demodulator (including some that cannot be realized in hardware), the digitized signal can be optimally filtered, demodulated, post-detection processed, and so forth.

The problems, of course, are in the implementation. The most critical element is the analog-to-digital (A/D) converter. Two samples per cycle of the highest frequency in the signal being digitized are necessary to provide adequate signal to the computer. The state of the art is moving forward almost daily, but limits still exist on the maximum frequency that can be digitized and the maximum resolution that can be provided.

The computer has a finite processing capability (however, this also is growing almost daily). This processing capability limits the signal data throughput. And complex software requires lots of storage and processing memory. Computer capability is a huge and interactive function of size, weight, power, and cost.

While the state of the art is moving in the right direction, making a full-frequency-band digital receiver is normally still impractical, so the system must translate a portion of the frequency range into the frequency band covered by the digital receiver. This frequency band is sometimes converted to a "zero IF" (the lower edge of the IF—intermediate frequency—band is

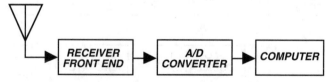

Figure 4.8   The digital receiver digitizes its IF passband and then applies appropriate software implemented filtering and demodulation functions to recover received signals.

near DC), or the IF is "subsampled." The subsampling of an IF occurs at a sample rate that is far lower than the IF frequency but equal to twice the highest modulation rate of the signal being digitized.

4.10 Receiver Systems

Virtually all modern EW and reconnaissance systems require more than one type of receiver to adequately perform their functions. The typical receiver system (or subsystem) configuration is shown in Figure 4.9. The inputs from one or more antennas are either power divided (if all receivers operate over the full frequency range) or multiplexed (if receivers operate over separate segments of the system's frequency range). In complex systems, the signal distribution involves a combination of both.

It is quite common for EW/reconnaissance systems with narrowband receiver requirements to task a single receiver (or set of receivers) with searching for new signals, which are then handed off to dedicated receivers. These dedicated receivers remain at their assigned frequencies with the assigned bandwidth and demodulation settings as long as required to completely analyze a signal, unless they are reassigned to a higher-priority signal.

Another common practice is to use a special processing receiver—typically more complex than the other receivers—to provide additional information about a signal being processed by one of several monitor receivers.

The following are examples of typical applications of multiple receiver types operating cooperatively in EW or reconnaissance systems. Not intended to cover all possible approaches, these examples illustrate several important receiver system issues.

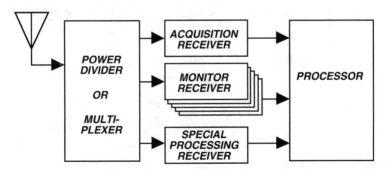

Figure 4.9 Virtually all modern EW and reconnaissance systems include multiple types of receivers to optimally handle different tasks.

4.10.1   Crystal Video and IFM Receivers Combined

Since an electronic support system, particularly a radar warning receiver (RWR), must very quickly determine all the parameters of each pulse it receives, it often uses crystal video and an instantaneous frequency measurement (IFM) receiver together (Figure 4.10). The crystal video receivers measure pulse amplitude, start time, and stop time, while the IFM unit measures the frequency of each pulse.

The multiplexer divides the input frequency range so that each crystal video channel covers a different band (for discussion, let's say 2–4 GHz, 4–6 GHz, and 6–8 GHz). The frequency converter folds each of these bands into a single frequency range for input to the IFM (e.g., 2–4 GHz). Thus, the IFM output is ambiguous (3 GHz, 5 GHz, and 7 GHz all look like 3 GHz to the IFM). However, the pulse analyzer receives the pulses from each separate band. By correlating the time at which the IFM measures a frequency with the time of the received pulses in each band, the pulse analyzer can resolve the IFM measurement ambiguity.

4.10.2   Receivers for Difficult Signals

When a reasonable number of "difficult-to-process" signals are expected in a wide-frequency-range signal environment, the solution is the selective use of a special receiver in the configuration shown in Figure 4.11. The best example of this is the modern RWR, which must handle a few CW or other challenging signals in a dense pulse environment. The individual-band

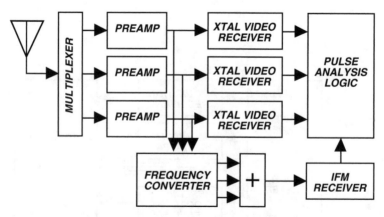

Figure 4.10   Crystal video and instantaneous frequency measurement receivers are often used in combination to provide pulse parameter data in high-density signal environments.

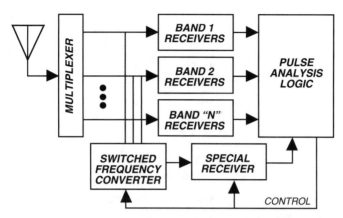

Figure 4.11 Modern RWRs use special receivers (digital, channelized, or superhetero-dyne) to identify and locate emitters with difficult modulations.

receivers are crystal video receivers, and the special receiver is a superhetero-dyne, channelized, or digital receiver. The signal analysis logic assigns the special receiver based on a combination of the data received from the regular-band receivers, foreknowledge of the expected environment, and perhaps an IFM configured as shown in Figure 4.10. If no other clues are available, the logic might simply cycle the special receiver through the whole frequency range following a prioritized search pattern.

In this case, the frequency converter would be as shown in Figure 4.12, and the special receiver would cover "Band 1." It is also possible to design the system so that more than one converted channel can be switched into the output, which would require that the frequency ambiguity be resolved. Note

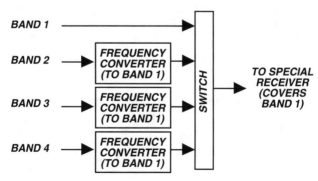

Figure 4.12 A multiple band converter is often used to heterodyne equal bandwidth portions of the system's full frequency range into one band for processing by a special receiver.

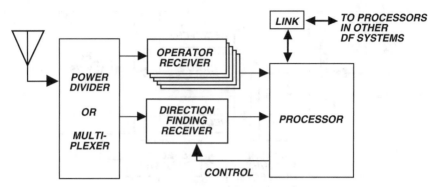

Figure 4.13   A typical communications-band direction-finding system shares a single DF receiver in each of multiple stations among several operators.

that the frequency converters are often designed so that each local oscillator serves more than one band converter; either a "high side" or "low side" conversion can be used in any of the converted bands. In high-side conversion, the LO is above the input band, and in the low-side conversion, it is below the input band. Depending on the input band and the LO frequency, the output band can be higher or lower in frequency and can be right side up or upside down (lowest input = highest output).

4.10.3 Special Receiver Time Shared by Several Operators

Figure 4.13 shows a common example of a special receiver providing a special function to many independent analysis receivers. In this case, a DF receiver is tasked by operators who are performing in-depth analysis on signals and need emitter-location information. Depending on the emitter-location technique employed (see Chapter 8), the DF receiver may require additional antennas and/or cooperative operation with one or more additional DF sites.

4.11 Receiver Sensitivity

Receiver sensitivity defines the minimum signal strength a receiver can receive and still do the job it is intended to do. *Sensitivity* is a power level, typically stated in dBm (usually a large negative number of dBm). It can also be stated in terms of the field strength (in microvolts per meter). Simply stated, if the output of the link equation (as defined in Chapter 2) is a "received power" equal to or greater than the receiver sensitivity, the link works—that is, the receiver is able to "adequately" extract the information contained in the transmitted signal. If the received power is below the sensitivity level, the information will be recovered at less than the specified quality.

4.11.1 Where Sensitivity Is Defined

Although not always followed, it is good practice to define the sensitivity of a receiving system at the output of the receiving antenna, as shown in Figure 4.14. If sensitivity is defined at this point, the gain of the receiving antenna (in dB) can be added to the signal power arriving at the receiving antenna (in dBm) to calculate the power into the receiving system. This means that the losses from any cable runs between the antenna and the receiver and the effects of any preamplifiers and power distribution networks are all considered in the calculation of receiver system sensitivity. Naturally, if you are buying a receiver from a manufacturer, the manufacturer's specifications will assume that there is *nothing* between the antenna and the receiver—so "receiver sensitivity" (as opposed to receiver system sensitivity) is defined at the receiver input.

Inherent in the above argument is that losses associated with any cables, connectors, and so forth, that are defined as part of an antenna (or antenna array) must be considered when defining the antenna's gain. These may seem trivial points, but "old hands" will tell you that misunderstandings in this area have caused many a loud argument when it came time to buy or sell a piece of equipment.

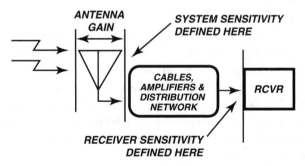

Figure 4.14 Receiver system sensitivity is defined at the output of the receiving antenna so that the minimum acceptable signal arriving at the antenna can be determined by the sum of the sensitivity and the antenna gain.

4.11.2 The Three Components of Sensitivity

There are three component parts of receiver sensitivity: the thermal noise level (called kTB), the receiver system noise figure, and the signal-to-noise ratio required to adequately recover the desired information from the received signal.

4.11.2.1 kTB

kTB (commonly used as though it were a real word) is actually the product of three values:

- k is Boltzmann's constant (1.38×10^{-23} joule/°K);
- T is the operating temperature in degrees Kelvin;
- B is the effective receiver bandwidth.

kTB defines the thermal noise power level in an ideal receiver. When the operating temperature is set at 290°K (a standard condition that is used to represent "room" temperature but is actually a cool 17°C or 63°F) and the receiver bandwidth is set at 1 MHz, and the result is converted to dBm, the approximate value of kTB is −114 dBm. This is often stated as:

$$kTB = -114 \text{ dBm/MHz}$$

From this "rule of thumb" number, the ideal thermal noise level in any receiver bandwidth can be quickly calculated. For example, if the receiver bandwidth is 100 kHz, kTB is −114 dBm − 10 dB = −124 dBm.

4.11.2.2 Noise Figure

If you don't buy your receiver from the mythical company an old professor used to call "the Ideal Receiver Store," it will add some extra noise to any signal it receives. The ratio of the noise present in the receiver bandwidth to that which would be present if only kTB were present is called the *noise figure*. Actually, that's not quite true—the noise figure is defined as the ratio (noise/kTB) of the noise that would have to be injected into the *input* of an ideal, noiseless receiver (or receiving system) to produce the noise that is actually present at its output (Figure 4.15). This same definition applies to the noise figure of an amplifier.

The noise figure of a receiver or an amplifier is specified by its manufacturer, but the determination of the system's noise figure is a little more complicated. Consider first the case of a very simple receiving system that has a single receiver connected to an antenna by a lossy cable (or any other

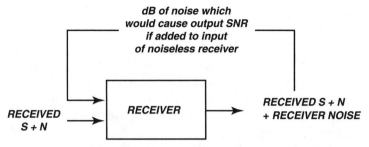

Figure 4.15 The noise figure of a receiver is the amount of thermal noise the receiver adds to a received signal, referenced to the receiver input.

passive device that has no gain—for example, a passive power divider). In this case, all losses between the antenna and the receiver are simply added to the receiver's noise figure to determine the system noise figure. For example, if there is a cable with 10 dB of loss between the antenna output and the input of a receiver with a 12-dB noise figure, the system noise figure is 22 dB.

Now, consider the noise figure of a receiving system that includes a preamplifier, as shown in Figure 4.16. The values L_1 (the loss between the antenna and the preamplifier in dB), G_P (the preamplifier gain in dB), N_P (the preamplifier noise figure in dB), L_2 (the loss between the preamplifier and the receiver in dB), and N_R (the noise figure of the receiver in dB) are defined variables. The noise figure (NF) of this system is determined by the formula:

$$NF = L_1 + N_P + D$$

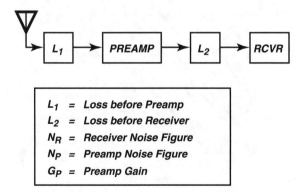

Figure 4.16 The noise figure of a receiving system can be reduced by the addition of a preamplifier.

where the values L_1 and N_P are just plugged in, and D is the degradation of the system noise figure by everything following the preamplifier. The value of D is determined from the graph in Figure 4.17. To use this chart, draw a vertical line from the value of the receiver noise figure (N_R) on the abscissa and a horizontal line from the value of the sum of the noise figure and gain of the preamplifier less the loss between the preamplifier and the receiver ($N_P + G_P - L_2$). These two lines cross at the degradation factor in dB. In the example drawn on the figure, where the receiver noise figure is 12 dB, the sum of the preamplifier gain and noise figure reduced by the loss to the receiver is 17 dB (for example, 15-dB gain, 5-dB noise figure, and 3-dB loss). The degradation is 1 dB. If the loss between the antenna and the preamplifier were 2 dB, the resulting system noise figure would be 2 dB + 5 dB + 1 dB = 8 dB.

4.11.2.3 Required Signal-to-Noise Ratio

The signal-to-noise ratio (SNR) required for the receiver to perform its task is highly dependent on the type of information carried by the signal, the type of signal modulation that carries that information, the type of processing that will be performed on the output of the receiver, and the ultimate use to which the signal information will be put. It is important to realize that the required

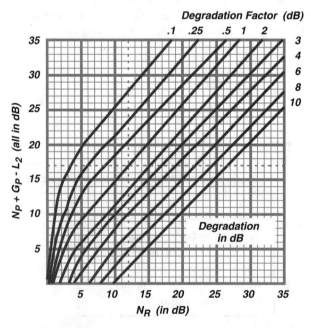

Figure 4.17 The degradation of system noise figure by everything following the preamplifier can be determined from this chart.

SNR that must be defined to determine the receiver sensitivity is the *predetection SNR,* called the *RF SNR* or sometimes the *carrier-to-noise ratio* (CNR). With some types of modulation, the SNR in the signals at the receiver output can be significantly greater than the RF SNR.

For example, if a receiving system has an effective bandwidth of 10 MHz and a system noise figure of 10 dB, and is designed to receive pulsed signals for automatic processing, its sensitivity is:

$$\text{kTB} + \text{noise figure} + \text{required SNR}$$
$$= (-114 \text{ dBm} + 10 \text{ dB}) + 10 \text{ dB} + 15 \text{ dB} = -79 \text{ dBm}$$

4.12 FM Sensitivity

Because of the nature of the modulation on frequency modulated (FM) signals, the sensitivity of an FM receiver is determined both by the received power level and the modulation characteristics. The received power must be great enough so that the SNR into the FM discriminator is adequate to recover the modulation. Once this "threshold" has been reached, the width of the frequency modulation determines an SNR improvement factor, which enhances the sensitivity.

Frequency-modulated signals represent the amplitude variations of modulating signals as changes in the transmitted frequency. (Figure 4.18 shows this for a sine-wave modulating signal.) The ratio of the maximum transmitted frequency excursion (from the frequency of the unmodulated carrier signal) to the maximum frequency of the modulating signal is called the modulation index, represented by the Greek letter β.

When properly demodulated, the output signal quality is increased above the RF SNR by a factor that is a function of the value of β—as long as the RF SNR is above a required threshold value.

4.12.1 FM Improvement Factor

The threshold RF SNR for a normal FM discriminator is approximately 12 dB. It is approximately 4 dB for a phase-locked–loop-type FM discriminator. Below these received signal RF SNR threshold values, the output SNR is severely degraded, but above these thresholds the output SNR is improved by an FM improvement factor defined by the following equation:

$$\text{IF}_{\text{FM}} \text{ (in dB)} = 5 + 20 \log_{10}\beta$$

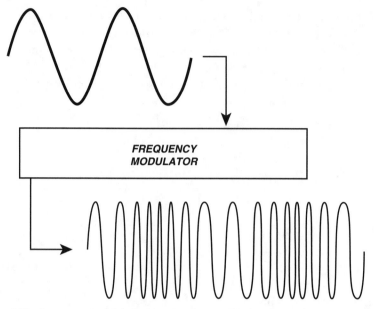

Figure 4.18 Frequency modulation carries the amplitude of a modulating signal as a change in transmitted frequency.

For example, if the receiver has a normal FM discriminator, if the received signal is strong enough to produce a 12-dB RF SNR, and if the received signal's modulation index is 4, then the FM improvement factor is:

$$\text{IF}_{\text{FM}} \text{ (in dB)} = 5 + 20 \log_{10}(4) = 5 + 12 = 17 \text{ dB}$$

The output SNR (directly in dB) is then:

$$\text{SNR} = \text{RF SNR} + \text{IF}_{\text{FM}} = 12 + 17 = 29 \text{ dB}$$

The achievement of this FM improvement factor depends on having the proper bandwidths as the signal moves through the receiver (i.e., the receiver designer must know what he or she is doing, which fortunately is almost always the case . . . almost).

As another example, assume that we need to output a 40-dB SNR (required for a "snow free" television picture). If the TV signal is transmitted as a frequency-modulated signal with a modulation index of 5, the required

RF SNR (which is necessary to determine the sensitivity of the receiver) is calculated as follows:

$$\text{IF}_{FM} \text{ (in dB)} = 5 + 20 \log_{10}(5) = 5 + 14 = 19 \text{ dB}$$

$$\text{Required RF SNR} = \text{Output SNR} - \text{IF}_{FM} = 40 - 19 = 21 \text{ dB}$$

4.13 Digital Sensitivity

The output quality of a digitized signal is a function of its modulation parameters. Too low an RF SNR will generate bit errors. (Yes, bit errors do degrade the quality of the signal, but they are normally considered separately from the quality of digitized analog signals and apply equally to digital signals that have never been analog—for example, an e-mail message.) The beauty of a digital signal is that it can be relayed again and again without a degradation in quality, as long as the RF SNR at each receiver is adequate to keep the bit errors at an acceptable level.

4.13.1 Output SNR

The "output SNR" for an analog signal that has been digitized is actually the signal-to-quantizing noise ratio (SQR). Consider Figure 4.19. When the original analog signal is digitized, then returned to analog form by a digital-to-analog converter at the output of the receiver, it will resemble the "reproduced digitized signal" shown in the figure. An appropriate filter will smooth out the sharp corners of the waveform, but the accuracy of the reproduction will not actually have improved, since only the digitized signal information was transmitted. A convenient expression for the SQR is in terms of the number of bits with which the signal amplitude is quantized:

$$\text{SQR (in dB)} = 5 + 3(2m - 1)$$

where m equals the number of bits per sample.

For example, the SQR for a signal that is digitized with 6 bits per sample is:

$$\text{SQR (in dB)} = 5 + 3(11) = 38 \text{ dB}$$

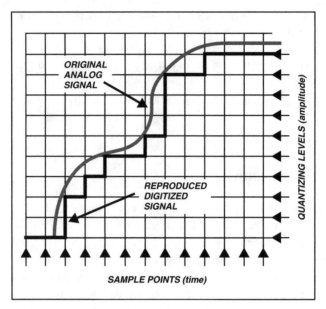

Figure 4.19   The accuracy of a reconverted analog signal which has been digitized is degraded by the quantizing process causing "quantizing noise."

4.13.2   Bit Error Rate

Any digitally formatted signal is transmitted as a series of "ones" and "zeros," which are modulated onto an RF carrier signal using some sort of modulation technique. Many specific modulation types are available, and each has its advantages and disadvantages—including the ratio of transmission bandwidth to digital data bit rate and bit error rate versus RF SNR performance. Under most circumstances, the various modulations will require an RF bandwidth-to-digital data ratio between 1 and 2 (i.e., 1 Mbps of data requires between 1 and 2 MHz of transmission bandwidth).

The bit error rate-versus-RF SNR performance is different for each type of modulation, but all tend to fall between the curves for a generic coherent phase-shift keyed (PSK) modulation and a noncoherent frequency-shift keyed (FSK) modulation, as shown in Figure 4.20. The bit error rate is the average number of incorrect bits divided by the number of bits transmitted. In the example shown in the figure, a digital signal that uses noncoherent FSK modulation and arrives at the receiver with an 11-dB RF SNR will have slightly less than a 10^{-3} bit error rate. If the modulation is coherent PSK, the bit error rate will be approximately 10^{-6}. Please note that the required

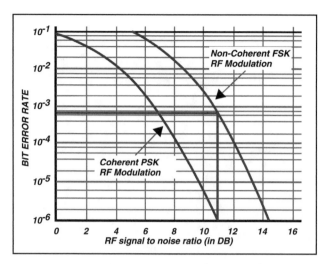

Figure 4.20 For any type of RF modulation used to carry digital data, the bit error rate of received signals is a function of the RF signal-to-noise ratio.

transmission accuracy for a digital data system is often specified in terms of the "word error rate" or "message error rate." Before we can use a chart like this to convert the error rate to a required RF SNR, we need to convert it to a bit error rate. For example, the bit error rate equals the message error rate divided by the number of bits in a standard message. If there are 1,000 bits in a standard message and only 1% of the messages can be incorrect (i.e., with one or more bit errors in the message), the bit error rate must be 10^{-5}.

5

EW Processing

There are three introductory comments that must be made relative to this coverage of processing. First, EW processing is a broad subject, and this chapter will not attempt to cover the entire field. Second, some of the subjects in other chapters could well be considered processing. This chapter will refer to those other chapters from time to time and tie them in with the flow of the current discussion. Third, the implementation of EW processing is changing almost daily because the capabilities of computer hardware are in a period of explosive growth. Therefore, the focus of this chapter is on what is done and why it is done—rather than the hardware or the specific software with which it is implemented.

5.1 Processing Tasks

EW is, by its nature, responsive to the threat signals present in its environment. Thus, from the beginning of modern EW in the early 1940s, it has been necessary to do some kind of processing to determine when and how to use the correct countermeasures. At first, there was complete dependence on skilled operators to determine which (if any) threat signals were present so that proper countermeasures could be used. Since human beings cannot directly detect radio-frequency signals, receivers detected the signals—which were then processed in some way for presentation in a form that operators could recognize.

As the signal environment became more complex, the radar-controlled weapons more lethal, and the timelines shorter, it was necessary to detect and identify threats automatically. Threat identification remains a primary EW processing task in almost all EW systems.

Emitter location is another task basic to EW operations. Emitter location (and direction finding) is covered in Chapter 8, so the techniques will not be covered here. However, the role of emitter location in higher-level processing functions is germane.

Since modern EW systems, particularly in airborne applications, must deal with many signals (including millions of pulses per second), the isolation of individual signals from the mass of received RF energy can be a critical processing function.

Modern EW systems are often highly integrated, including multiple sensors and multiple countermeasures. All of these system assets must be controlled and coordinated. We have already dealt with the control of multiple receivers in the search role (in Chapter 4), but we will cover some more specific selection criteria in certain EW applications.

The processing functions directly associated with jamming can be implied from the descriptions of the jamming techniques described in Chapter 9. Therefore, only the processing associated with jammer control will be considered here.

Table 5.1 is a top-level overview of the major types of EW processing and their roles in the EW mission. This is an admittedly arbitrary division of this very complex area, and EW processing professionals (like professionals in

Table 5.1
EW Processing Tasks

Processing Task	Role in EW Mission
Threat identification	Determines the type of emitter from signal parameters
Signal association	Assigns signal components to signals to support threat identification
Emitter identification	Identifies individual emitter (vs. emitter type)
Emitter location	Determines direction of arrival of signal or emitter location
Sensor control	Assigns sensor assets of an EW system based on analysis of data
Countermeasure control	Generates control inputs for countermeasures in an integrated EW system based on received signal data
Sensor cueing	Reduces the parametric search volume for narrow aperture assets
Man-machine interface	Reads control inputs and generates displays
Data fusion	Combines data from multiple sensors or systems to generate an electronic order of battle

any field) disagree on any general characterization of their field. The purpose of this table is to generate a logical structure within which we can discuss EW processing.

5.1.1 RF Threat Identification

Let's start with the problem of identifying threats from the parameters of received RF signals. In general, the parameters of a threat signal include:

- Effective radiated power;
- Antenna pattern;
- Antenna scan type(s);
- Antenna scan rate(s);
- Transmitted frequency;
- Type(s) of modulation;
- Modulation parameters.

When these signals arrive at a receiver, they are characterized somewhat differently. The received signal parameters are:

- Received signal strength;
- Received frequency;
- Observed antenna scan;
- Type of modulation;
- Modulation parameters.

Some parameters are relatively easy to measure, but some are difficult—requiring the use of special assets. Because threat identification in EW is typically a real-time process, the order in which parameters are analyzed must be carefully considered.

5.1.2 Logic Flow in Threat Identification

Threat identification in modern systems is highly complex—because there can be many threats present, and because the threat parameters are becoming more complex. In general, we must know what type of threat is present, the location of the threat and the operating mode of the threat. For RF-guided threats, all three of these items are usually determined from the received RF signals.

Here are three useful generalizations about threat-identification logic flow:

- The easiest analysis tasks are done first. These are usually the tasks that require only the use of wideband assets and/or very short signal intercepts.
- Signal data from early, easy analysis is removed, allowing more complex analysis to be performed on depopulated data.
- The analysis is terminated as soon as all necessary ambiguities have been resolved.

As an example, consider a radar warning receiver (RWR) operating against pulsed emitters. The signal parameters we must analyze are:

- Pulse width;
- Frequency;
- Pulse-repetition interval;
- Antenna scan.

These received signal parameters are illustrated in Figure 5.1.

As shown in Figure 5.2, the RWR would first attempt to determine threat type from parameters present in each pulse (frequency and pulse width). If the type of threat signal can be identified from just those two parameters, the processor will stop its analysis and report out the threat ID.

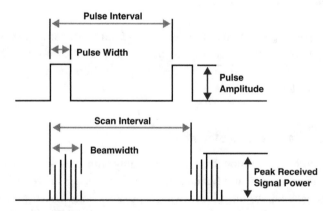

Figure 5.1 The pulse and scan parameters of a radar signal are analyzed to determine the type of radar producing them.

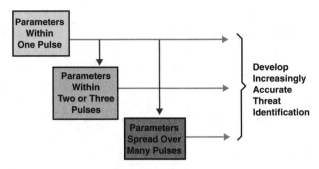

Figure 5.2 Processing for threat identification is normally performed in the order of increasing time required for data collection.

Next, the RWR will consider pulse-repetition interval, because that requires only the determination of the intervals between as few as two pulses. Unfortunately, there can be complications. If there are multiple pulse trains present, the pulses must be sorted into the individual signals. Also, the pulse train might not have just a simple pulse-repetition interval; it might be staggered or jittered. However, the fact remains that the analysis of the pulse intervals is the second-easiest task and, thus, would be handled second. If this yields an identification, the processor will stop here.

Finally, the RWR will consider antenna scan. Since this involves the analysis of the relative amplitudes of a long series of pulses, it requires the consideration of many sequential pulses that have already been associated with individual signals. This is the most difficult task because it is the most time-consuming. In fact, the interval between received antenna beams may be of the same order of magnitude as the total specified time for the RWR to complete its analysis and report out the threat ID.

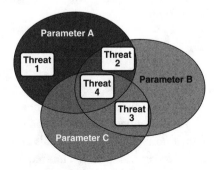

Figure 5.3 EW processors usually evaluate only enough data to resolve ambiguities among the threat types they are designed to identify.

A hypothetical threat identification case is shown in Figure 5.3. Three signal parameters are measured, and there are four possible types of threats. Threat 1 is easiest to identify because it can be unambiguously identified by the measured value of parameter A. Threats 2 and 3 both require the determination of values for two parameters to resolve their ambiguity, so they require more analysis effort than threat 1. Threat 4 can only be unambiguously identified by the determination of values for all three parameters.

5.2 Determining Values of Parameters

The first step in the analysis of threat signals is to measure the received signal parameters. To understand the measurement mechanisms, it is worthwhile to consider the way these measurements were made before computers were available to RWRs. Each parameter measurement circuit was built from discrete components and could only do a single task. The computers in modern systems do the same jobs, but with more dignity (or at least a lot more efficiently).

5.2.1 Pulse Width

When a pulse is passed through a high-pass filter, the result is a positive spike at the leading edge and a negative spike at the trailing edge, as shown in Figure 5.4. By using the positive spike to start a counter and the negative spike to stop the count, it was possible to very accurately measure the pulse width. A second approach is shown in Figure 5.5. The pulsed signal can be digitized at a high sample rate and analysis made to determine the pulse width. This approach also provides detailed information about the shape of the pulse. This approach is required in systems that measure rise time, overshoot, and so forth, in addition to the pulse width.

5.2.2 Frequency

In early RWRs, which used crystal video receivers, the frequency of received signals could only be determined by dividing the input into frequency ranges with filters and placing a crystal video receiver on each filter output. The frequency of pulsed or continuous wave (CW) signals could also be measured by tuning a narrowband receiver to a signal. The frequency of the signal was the frequency to which the receiver was tuned.

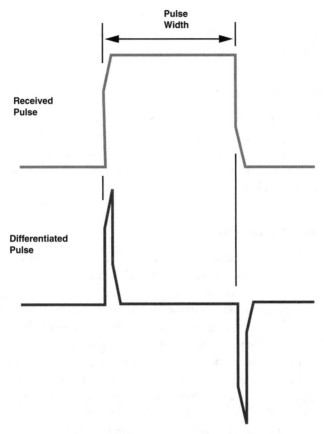

Figure 5.4 By starting and stopping a counter with the leading and trailing edge spikes, the pulse width can be measured with great accuracy.

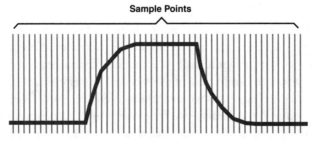

Figure 5.5 If the pulse waveform is sampled at a high rate, the complete shape of the pulse can be captured digitally.

With the advent of practical instantaneous frequency measurement (IFM) receivers—and computers to collect the data—the frequency of each pulse could be measured and stored.

5.2.3 Direction of Arrival

The direction of arrival (DOA) of each pulse is measured using one of several direction-finding approaches that are described in Chapter 8. Low accuracy DOA measurement was (and is) done using amplitude-comparison direction finding and high-accuracy DOA measurement was (and is) performed using an interferometric approach.

5.2.4 Pulse Repetition Interval

In the good (but hard) old days, the pulse repetition interval (PRI) of pulsed signals was measured using what was called a "digital filter." This was a device designed to detect the presence of a specific pulse interval. The digital filter opened an accept gate a fixed amount of time after a received pulse. If a pulse occurred when the gate was open, it would look for another pulse at the same interval. When a sufficient number of qualifying pulses had been received, the presence of a signal with the specified PRI could be determined. It was necessary to have one digital filter circuit per threat PRI, and multiples to handle staggered pulse trains. One of the charms of this approach was that the pulses from a single signal could thus be "deinterleaved" from the combined pulse trains of many signals in a wideband receiver.

Now, of course, a computer can collect the times of arrival of the leading edges of a large number of pulses and determine multiple PRIs and staggered PRIs mathematically.

5.2.5 Antenna Scan

Early RWRs had to determine the beamwidth of a threat emitter by setting a threshold and measuring the number of sequential pulses received above that threshold as shown in Figure 5.6. As a threat antenna beam scans past the receiver's location, the amplitude of the received pulses varies as shown in the figure. Thus, counting pulses worked unless there were other signals present during the count. Now, because we have better tools to deinterleave signals, the pulses from a single signal can often be isolated, and the shape of the pulse amplitude history curve calculated.

A histogram of DOA versus received power can be used to determine the type of antenna scan. Figure 5.7 shows the (highly unlikely) situation in

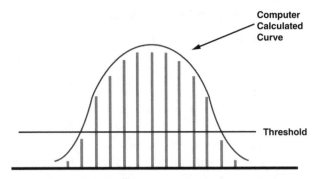

Figure 5.6 The antenna beamwidth of a pulsed threat signal can be sensed by counting pulses above a threshold or by analyzing the shape of the pulse amplitude history.

which three signals with different types of antenna scans are located along one DOA. The vertical axis is the number of hits (or pulses) received at that power level. If you will think about the time versus received power history for various types of scans, you will be able to see that the shapes shown differentiate among the three scan types.

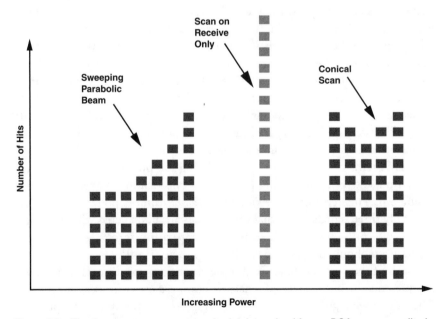

Figure 5.7 The threat antenna scan can also be determined from a DOA versus amplitude histogram. This drawing shows three histograms, all from the same DOA.

5.2.6 Receiving Pulses in the Presence of CW

Earlier in this chapter, we discussed an RWR that operated in an ideal world without any CW or very high duty cycle pulsed (primarily pulse Doppler) signals. The logarithmic response of a wideband receiver (for example, a crystal video receiver) will be distorted when CW signals are present along with pulses. When very high duty cycle pulse signals are present, their pulses overlap the low duty cycle pulses causing the same problem. Since accurate amplitude measurements are required in order to determine DOA, the CW signal precludes proper system operation against pulses. It should also be noted that an IFM receiver is a wideband receiver that can operate on only one signal at a time. The answer is to filter out the CW signal with a band stop filter. The wideband receiver can then "see" pulses elsewhere in its frequency range while the narrowband receiver handles the CW (or pulse Doppler) signal.

5.3 Deinterleaving

It has been noted that increased receiver bandwidth allows increased probability of intercept—with a wide-open receiver providing the ultimate in frequency-versus-time performance. Likewise, the probability of intercept is enhanced by increasing the instantaneous angular coverage—with 360° coverage being required in many EW systems. One of the problems with increasing bandwidth and/or instantaneous angular coverage, particularly in a dense environment, is that we are much more likely to have to deal with multiple, simultaneous signals. In this section, we will deal with the case of multiple pulsed signals received in the same receiver channel at the same time. This case deliberately ignores very high duty cycle signals, which we assume have been (somehow) removed from the signal set before this analysis begins.

Deinterleaving is the process of isolating the pulses of a single emitter from a pulse stream containing pulses from two or more signals. Consider Figure 5.8. This represents the video from a very simple pulse environment with only three signals. Note that these signals are all depicted with very high duty cycle (pulse width divided by pulse-repetition interval). You would expect normal pulsed signals to be about 0.1% duty cycle.

All of these signals have fixed pulse-repetition frequency (PRF). Signal B represents the beam shape of a narrow-beam radar as it passes the receiver. The other two signals are shown as having constant amplitude—perhaps because our sample falls within the emitters' beams.

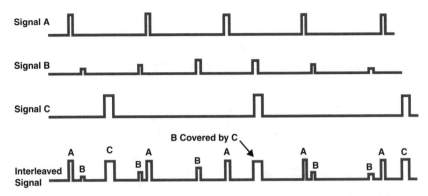

Figure 5.8 If multiple pulsed signals are received in the same receiver channel, it is necessary to perform deinterleaving to isolate the individual signals.

The "interleaved signal" line of the figure represents the combination of all three signals as it would appear within a wideband receiver. Each of the pulses is labeled as to its signal. When deinterleaved, the three pulse trains are separated into their respective individual signals. This allows further processing to take place.

5.3.1 Pulse on Pulse

Note that the second pulse in signal C covers the fourth pulse of signal B. This is called the "pulse on pulse" or "POP" problem. If the system sees only one pulse in this location, it will eliminate that pulse from one of the deinterleaved signals. Depending on the number of pulses thus eliminated and on the nature of the signal identification processing that follows, this may negatively effect the system performance.

Figure 5.9 takes a closer look at the two overlapping pulses. Note that the amplitude and duration of each pulse are present in the combined video signal. Thus, if the system processing has adequate resolution to measure these values, both pulses can be associated with their proper signals. It should be noted, however, that the receiver supplying the video waveform to the processing must have adequate bandwidth to pass the combined video with adequate fidelity to allow these measurements to be made.

5.3.2 Deinterleaving Tools

The deinterleaving process involves the use of everything we know about each received pulse—which, of course, depends on the configuration of the

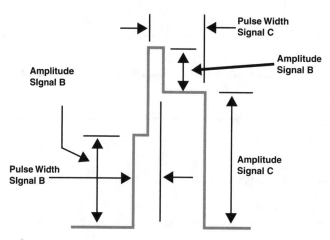

Figure 5.9   A detailed look at the video waveform of two overlapping pulses shows that the amplitude and width of each pulse are recoverable. The combined amplitude of the pulses during the overlap period is shown less than the sum of the two amplitudes because the logarithmic video output of a typical EW receiver would compress it.

receiving system. Table 5.2 shows what the system knows about each pulse based on the type of receiver in which the signal is intercepted. If a receiving system has some combination of these receiver assets, the processor will have the associated information on each pulse to which that receiver asset is

Table 5.2
Information Available for Each Pulse Versus Receiver Type

Type of Receiver or Subsystem	Information Measured on Each Pulse
Crystal video receiver	Pulse width, signal strength, time of arrival, and amplitude vs. time
Monopulse DF system	Direction of arrival
IFM receiver	RF frequency
Receiver with AM and FM discriminator	Pulse width, signal strength, time of arrival, RF frequency, and amplitude & frequency vs. time
Digital receiver	Pulse width, signal strength, time of arrival, RF frequency, and FM or digital modulation on pulse
Channelized receiver	Pulse width, signal strength, time of arrival, frequency (by channel only)

applied. However, since the system may time-share some of its receiver assets among frequency bands, it is not usually safe to assume that all of the information will be available for every pulse.

Note that the types of receivers used in EW were described in Chapter 4.

Obviously, deinterleaving is easier if signals can be separated by clearly identifying each pulse—then sorting them into signals. This requires not only the measurement of parameters, but adequate resolution to differentiate between the signals for each parameter used.

In the early days of modern RWR development, it was typical to have only crystal video receivers in a monopulse direction-finding system. Since received-pulse amplitude can vary—from pulse to pulse—as a threat emitter's antenna scans past the receiver, only the timing and direction of arrival could be used. However, the direction-of-arrival-measurement output was relatively inaccurate and varied as a function of variations in the receiving-antenna gain patterns. Thus, direction of arrival was not a dependable parameter. This meant that the time of arrival of pulses was the only practical method for deinterleaving pulses unless they could be separated by pulse width.

The original technique for pulse-interval deinterleaving was described in the previous section, but it should be noted that this method was most effective against fixed PRF signals. Staggered pulse trains could be identified if there was one "digital filter" per phase of the stagger, but jittered pulse trains were a distinct problem. Using computer processing to identify pulse intervals from the times of arrival of a series of pulses simplified the staggered-pulse processing, but the deinterleaving of jittered pulse trains remains quite difficult unless the individual pulses can be identified by some means. This process is greatly enhanced by identifying the pulses of simple pulse trains and depopulating them from the data before processing the more complex pulse trains.

When IFM receivers became available, pulse-by-pulse frequency could be measured. This provided a powerful tool for sorting pulses into frequency bins which could normally be associated with individual signals—a powerful deinterleaving tool if adequate processing power and memory are available. This technique may break down for threat signals which have pulse-to-pulse frequency agility. Again, if all the pulses from simple pulse trains can be depopulated from the data before handling these complex signals, it may be practical to associate frequency-varying pulses—unless there are multiple frequency-agile radars of the same type in the receiver at the same time.

If a high-accuracy direction-finding system is available, and if it provides stable direction-of-arrival data on a pulse-by-pulse basis, the pulses can be deinterleaved by direction of arrival. In most circumstances, this would be a highly desirable deinterleaving scheme, because it would work even with signals that have very complex modulation (for example, both pulse and frequency agility). Once the pulses of a single signal are isolated, statistical analysis can be performed to derive the necessary information from the modulation.

5.3.3 Digital Receivers

As digital receivers become more available and more capable, all of the techniques used in the old types of systems will be available in software. As long as a signal can be digitized with adequate fidelity, you can perform just about any type of process using software. This includes adaptive demodulation, filtering, parameter extraction, and so forth. However, "adequate fidelity" is a serious qualifier. The digitizing limitations are the number of bits per sample (which limits the dynamic range over which processing can be done) and the digitizing rate (which limits the time fidelity of processing). Both of these limits are being tested by new technology developments on an almost daily basis—so watch this technology carefully.

5.4 Operator Interface

One of the challenging electronic warfare processing tasks is operator interface (also called man-machine interface, or MMI). The system must accept commands from operators and provide data to them. The challenge is to make EW systems "user friendly," which means to accept commands from the operator in the form that is most intuitive to the operator, and to present information to the operator in the most directly usable form possible or practical. This simple statement can have significant impact when applied. We will consider two specific EW system applications to illustrate the issues. These are an integrated aircraft EW suite—and a tactical emitter location system netted with other, remote direction-finding systems. For each of these examples, we will characterize the commands and data involved, and will discuss the history of display developments, the current common approaches, anticipated trends, and timing issues.

5.4.1 In General (Computers Versus Humans)

The general problem is that computers and people have totally different approaches to the input/output (I/O) of information (as shown in Figures 5.10 and 5.11). Computers like their I/O information to be compatible with

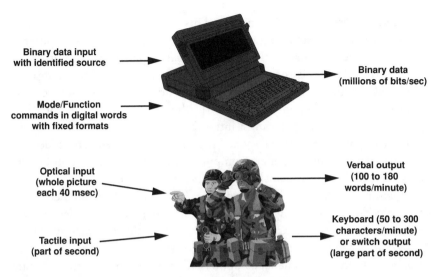

Figure 5.10   The form in which information is input and output is significantly different for computers and humans. The data rates (but not necessarily the effective information rate) are also significantly different.

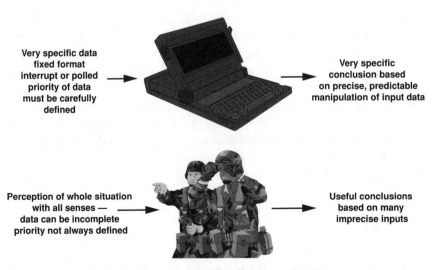

Figure 5.11   The ways in which humans and computers process information are completely different. Humans can use less specific information to form situationally adaptive conclusions.

the internal working of the computer. This means control inputs must be available (in simple, unambiguous, digital format) when the computer is ready to use the information. It also means that displayed data is output (in digital form) as soon as the computer has made its calculation. Computer I/O speeds are up to millions of bits per second. Computer inputs can be either polled (i.e., the computer looks for the data when it is required) or interrupts (i.e., the computer must interrupt some part of its work to accept the input). Computers "prefer" polled inputs, because interrupts reduce the computer's processing efficiency. Computers generate the actual output data in digital form and like to clock it out at the full I/O rate.

Computers are very "black and white" in their I/O requirements. Try typing a period when a computer wants a comma or an uppercase letter when the computer wants lowercase if you need to prove this point. Input values are accepted as precisely correct and output values are generated at their full available resolution. In general, computers accept all properly formatted data sent to them—unless the peak data rate is too high or the average rate exceeds the processing throughput rate.

We people, on the other hand, like our I/O integrated with our other activities. We communicate in complex and sometimes contradictory human languages; words have different meanings depending on the context and the time and place at which they are used. Although we can receive information through our eyes, ears, or sense of touch, we get about 90% of our information visually. We accept information more efficiently (and remember it longer) if it is received simultaneously via two channels (visual and aural, and visual and tactile or aural and tactile).

People can accept a vast amount of information at an incredible rate if it is presented in context and is relevant to our experience. On the other hand, we accept random or abstract information very slowly and must relate new information to some familiar frame of reference before we can use it. Another characteristic of human information utilization is that we can accept multiple inputs of less than 100% correct or complete data and compile it into correct information.

The way that these computer/human information-handling differences are solved is the basis of the discussions of the two operator-interface examples chosen.

5.4.2   Operator Interface in the Integrated Aircraft EW Suite

When we started upgrading combat aircraft EW capabilities early in the Vietnam War, almost all of the EW systems and subsystems had their own controls and indicators. The operator had to spend a significant amount of training time just learning "knobology." Data from the systems had to be absorbed and interpreted by the operator, who then had to manually initiate the appropriate countermeasures. For example, the electronic warfare operator (EWO) position in the B-52D had 34 separate panels (plus some equipment located behind his seat). These panels contained a total of over 200 knobs and switches with close to 1,000 possible switch positions plus proportional analog adjustments.

The controls in these early EW systems allowed (and required) the operator to directly modify specific performance parameters of the equipment. The status displays showed specific equipment operating conditions, while received-signal displays showed the parametric details of individual signals.

Perhaps the most commonly used enemy-signal-detection device was the radar warning receiver. Its displays included a vector scope and a panel with lighted pushbutton switches. Figure 5.12 shows the vector scope as it

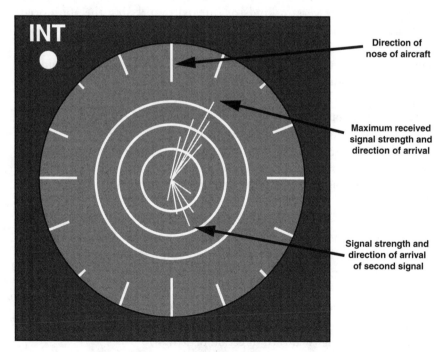

Figure 5.12   Early RWR displays (employed early in the Vietnam War) included a vector scope which showed the direction of arrival of individual pulses. The operator's eye "integrated" the displayed information to determine the maximum length of the strobe and its angle on the scope. The length of the strobe was proportional to the received signal strength, and its position on the scope showed the direction of arrival relative to the nose of the aircraft.

was used in the AN/APR-25 RWR. The vector scope was mounted on the instrument panel of most combat aircraft. Received signals were displayed as strobes on this display. The top of the display represented the nose of the aircraft, and the strobes indicated the relative direction of arrival of threat signals. Although the strobes were not stable, they varied randomly about a mean direction, so the operator could readily determine the direction of arrival within a few degrees. The length of the strobe represented the received signal strength. This signal strength indicated the approximate range to the transmitter. This system used the multiple antenna amplitude comparison DF technique described in Chapter 8, and the way that received signal strength varies with range to the transmitter is described in Chapter 2. The RWR also had circuitry that determined the type of threat signal(s) present using the earliest techniques described in the last three columns. The panel of lighted switches indicated (by which switches were lighted) the types of threats present. The operator could change the system's operating mode (for example, to ignore certain kinds of threats) by pressing the appropriate switches.

An additional signal-recognition aid available to operators was an audio signal generated by stretching received pulses so that the operator could hear the pulse-repetition frequency. As the threat antenna scanned the aircraft, the amplitude of received signals would vary, creating unique sounds which operators were trained to recognize. (The sound of the SA-2, for instance, was usually described as "like a rattlesnake.")

If there were multiple threats present, it was sometimes difficult to determine which type of threat was in which location. This type of display was effective when used by highly skilled operators in relatively low-threat densities.

Figure 5.13 shows how the vector-scope strobes were generated. There was one strobe per received pulse, generated by supplying current ramps to the magnetic-deflection coils surrounding the cathode-ray tube which formed the display. The direction and amplitude of the strobe were caused by the vector sum of the peak current values in the four deflection coils.

As the war continued, a second generation of processors replaced the pulse-by-pulse strobes with coded strobes which gave some information about the type of signal at each direction of arrival.

Thus far, the processing was largely done in specialized analog and digital hardware.

At the end of the Vietnam War, so-called "digital displays" were incorporated into RWRs. A typical early digital display is shown in Figure 5.14. Once the type of threat was identified, a computer generated a symbol to

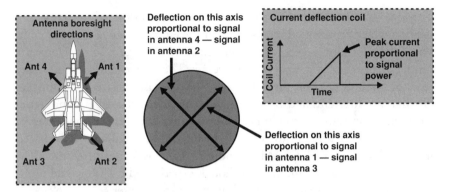

Figure 5.13 The signal currents to the magnetic deflection coils in the vector scope were proportional to the signals received by the four antennas on the aircraft. A current ramp was input to each coil as each pulse was received. This formed the strobe on the scope face.

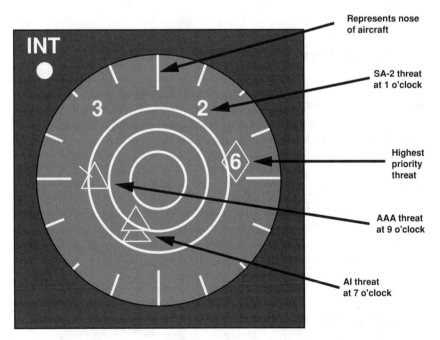

Figure 5.14 Third-generation RWR displays (at the end of the Vietnam War) provided symbols to identify threat type. The symbols were positioned on the screen to indicate the location of the emitter. Typical symbols are shown; they are completely selectable by the manager of the design program.

represent that threat type. The symbol was placed on the vector scope at a location which represented the location of the emitter relative to the aircraft. The aircraft location is generally at the center of the screen, so the closer the emitter, the closer the symbol is to the center. There were (and are still) many types of symbols used on these displays. For this case surface-to-air missiles (SAMs) are differentiated by type, while anti-aircraft artillery and aerial interceptors are shown as graphic symbols. Also, there are various types of symbol modifiers used on the display. In Figure 5.14, a diamond is placed around the 6 (for an SA-6 SAM) to indicate that this is considered the highest priority threat at the moment (although experienced EWOs would probably be more than casually interested in that enemy fighter at seven o'clock). Symbol modifiers are used to indicate the modes of the threat emitters (for example, tracking or launch mode) or to indicate which threat signals are being jammed.

During this time period, the controls for the jammers were still separate, but you can see the beginning of the integration of the functions (modification of symbols on the vector scope).

5.5 Modern Aircraft Operator Interface

As the threat environment has become denser (and more lethal), it has become necessary to transfer more information to the operator in less time. To do this, and and to enable the operator to take decisive action within the shortening allowable response times, the information must be presented "situationally." Although it comes as a shock to some engineers, a fighter pilot, especially when upside down, pulling 6 Gs, and figuring out how to stay alive for the next 5 seconds, is not interested in working differential equations to understand the tactical situation. The computer providing that information must speak fluent "fighter pilot."

Modern EW displays integrate the tactical picture for the operator and present the information in a quickly usable form. We will discuss modern aircraft displays, and then ground tactical displays.

5.5.1 Pictorial Format Displays

The figures in this section are from a USAF study (AFWAL-TR-87-3047 Final Report). Figure 5.15 is the general instrument-panel layout—basically the cockpit layout used in the F/A-18 and several other aircraft. As shown in the figure, there are five pictorial displays: the head-up display (HUD), the vertical-situation display (VSD), the horizontal-situation display (HSD), and

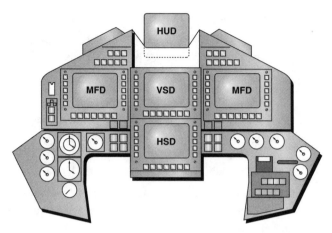

Figure 5.15   The modern instrument panel includes five displays: an HUD, a VSD, an HSD, and two MFDs.

two multiple-function displays (MFD). These types of displays can be used for any aircrew station, but since some aircraft have only one aircrew member (the pilot), the following discussion will focus on the pilot's displays.

5.5.2 Head-Up Display

The main reason for the use of a HUD is the large part of a second that it takes a pilot to move from "inside the cockpit" to "outside the cockpit." The eyes of a pilot who is "inside the cockpit" are focused at short range, and the pilot's mind is oriented to the artificial way that the instruments present the world. When the pilot is "outside the cockpit," his eyes are focused at long range, and his mind is oriented to the colors, brightness, angular movement, and moving objects of the real world.

The HUD allows the pilot to get some of the critical information available inside the cockpit without "moving inside." The HUD display is a cathode-ray tube which projects through a complex prism onto a hologram placed on a piece of glass directly in the pilot's field of view. In areas not containing data, the HUD is transparent. Figure 5.16 shows the basic HUD symbology. The airspeed, heading, and altitude are shown in standard positions. In the center of the display, there is an "ownship" symbol as a reference for the other data. The pathway symbol shows the pilot where to fly to avoid threats or terrain. Symbols for active threats can be displayed in the area below the "zero-pitch reference line."

In the air-to-air-combat mode, special displays relative to the lethal zones of ownship and enemy weapons can be placed on the HUD.

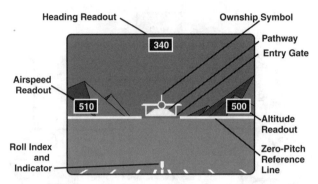

Figure 5.16 The HUD displays the information that is available "inside the cockpit" and presents it to the pilot directly in his line of vision when looking outside the cockpit.

5.5.3 Vertical-Situation Display

Figure 5.17 shows the vertical-situation display in the "ground mode." This is a view of the perspective from behind the aircraft. Note that airspeed, heading, and altitude are displayed in the same locations as on the HUD. Terrain features are shown as they would appear if they could be seen directly. The most striking aspects of this display are the lethality zones of the threats that are detected by the aircraft's RWR. The RWR determines the type of threat and its location. From previous electronic-intelligence analysis, the three-dimensional lethal zone of each type of threat is known. Thus, the computer can present each weapon's lethal zone situationally, allowing the pilot to avoid them. In most descriptions of this type of display, the lethal zones are

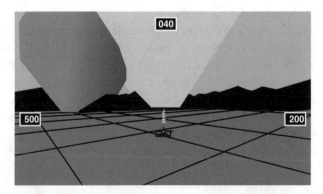

Figure 5.17 The VSD shows the situation around the aircraft as it would be seen from a position behind the aircraft. This is the ground-mode VSD.

divided into the full lethal area (usually depicted in yellow) and the part of the lethal zone in which no escape is possible (usually depicted in red). The lethal zones of weapons detected by other aircraft, but not by the ownship, can also be displayed and identified as "prebriefed."

There is also an "air mode" for the VSD in which airborne and ground-based threats are placed where they would be seen if the pilot could view the situation from the display's perspective.

5.5.4 Horizontal-Situation Display

The horizontal-situation display (ground mode), shown in Figure 5.18, is a view of the aircraft and its surroundings from above. The ownship symbol is in the center, and the digital bearing is at the top. Since the pilot can adjust the scale of the display, the current scale factor is shown at the bottom left. The flight path is shown as a series of lines and waypoints. Threats are shown as lethal areas (higher lethality in the center) at the aircraft's current operating altitude. Terrain is indicated as areas that extend above the aircraft's current altitude. Tactical situation elements are also shown on the display. For example, the forward line of troops (FLOT) is shown as a line with triangles pointing toward the enemy.

The HSD air mode shows the elements important to air-to-air combat. The lethal range of the ownship air-to-air weapons is shown in front of the ownship symbol. Enemy aircraft, along with their weapons' lethal zones, are shown in appropriate colors (usually red).

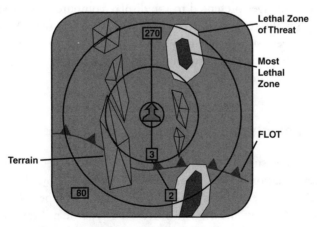

Figure 5.18 The HSD shows the flight path, along with terrain features and threat-lethality envelopes at the aircraft's current altitude.

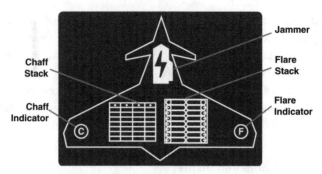

Figure 5.19   A typical MPD (one of dozens used) shows the status of the aircraft's countermeasures.

5.5.5   Multiple-Purpose Displays

The multiple-purpose displays (MPD) are used to display less immediate information to the aircrew in pictorial formats. Examples of this type of information are engine thrust, fuel status, hydraulic system status, weapons status, and so forth. Figure 5.19 shows the countermeasures status in a pictorial format. Dozens of such displays can be called up by the aircrew as required. Some displays will be automatically presented because they include information which is becoming critical—for example, "Bingo fuel," engine out, engine fire, etc.

5.5.6   Challenges

One of the challenges presented to the display computers by these types of displays is to make them consistent with what the pilot sees. Aircraft pitch and yaw relatively slowly, but they can roll at extremely high rates. Since these displays are processing intensive, care must be taken to update the displays at the aircraft-roll rate.

5.6   Operator Interface in Tactical ESM Systems

Tactical electronic support measures (ESM) systems are designed to give the commander situational awareness. The ESM systems allow the determination of the enemy's electronic order of battle (i.e., the types of the enemy's transmitters and their locations). Since each type of military asset has a unique combination of emitters, knowledge of the types of emitters present and their relative and absolute locations allows analysts to determine the make-up and location of the enemy's forces. It may even be possible to determine the enemy's intentions from the electronic order of battle.

5.6.1 Operator Functions

Once emitter locations have been determined, they can be passed to higher-level analysis centers electronically; however, the operator(s) of an ESM system need to be able to evaluate data in order to determine valid emitter locations and are assisted by special displays. A unique characteristic of tactical ESM systems operating against ground forces is that a single system can seldom determine the location of an enemy emitter. DOA must be measured from multiple sites at known locations, as shown in Figure 5.20. The emitter location is then determined by triangulation. The measured emitter location is, of course, the point at which two DOA lines cross.

In an ideal (flat and empty) world, two DOA cuts would be sufficient to determine the location of the emitter. In the real world, however, reflections from terrain cause multipath signals. Terrain can also block the line-of-sight path to a receiver. Also, other transmitters at the same frequency may cause one or more of the direction-finding (DF) receivers to give false readings. These three factors will cause each line of bearing to be less than absolutely accurate.

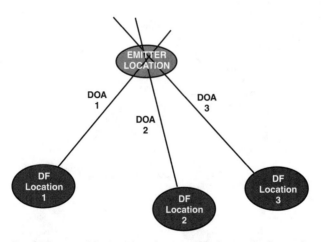

Figure 5.20   Establishment of the location of a ground emitter typically requires that direction of arrival be determined from a minimum of two known locations. Three DF measurements allow the accuracy of the location to be evaluated.

5.6.2 Real-World Triangulation

When there are three DF receivers, there are three triangulation points. In the real world, these points will not be colocated (as shown in Figure 5.21). The greater the multipath and interference effects, the greater the spread among the calculated line-intercept points. If several DF measurements are made, the statistical variation of these location points can be used to calculate a quality factor for the emitter location; the smaller the statistical spread, the higher the quality factor. An operator who can see the bearing lines and knows the area's terrain can eliminate obviously bad lines of bearing from the calculation, allowing the computer to calculate the most accurate emitter location possible. Thus, it is important that the operator's display associate the DF-receiver locations, the lines of bearing, and the tactical situation with terrain features.

Many years ago, when the density of emitters was fairly low and the tactical situation was usually not very fluid, it was possible for DF operators to verbally report their DOA readings to the analysis center. An analyst would plot the DF site locations on a tactical map and draw the reported lines of bearing. Then, the analyst could read the triangulation coordinates from the map. Computer-generated displays were an obvious improvement, particularly as the signal density increased and tactics became more mobile.

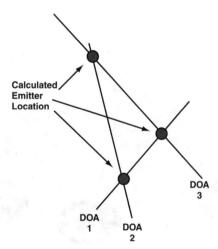

Figure 5.21   The three intersections of the three lines of bearing in Figure 5.20 would ideally be colocated. The spread of their locations over several sets of measurements gives a measure of the accuracy of the location.

5.6.3 Computer-Generated Displays

Early computer displays were as shown in Figure 5.22. A line drawing of important terrain features was made, and then the tactical situation was superimposed. The DF-receiver locations and lines of bearing were drawn on by the system. The data at the left of the screen allowed the operator to associate the triangulation points with signal frequencies (and any other signal information available). From this display, the operator could remove obviously incorrect lines of bearing and zoom in for close-up analysis of triangulation points. Once the data was edited, the system could report the locations and associate them with other known signal data.

Obviously, it would have been desirable to combine the computer-generated data with a tactical map on a single display, but digital maps were not yet available. One of the solutions was to position a video camera over a tactical map to create an electronic image. Then, the video map could be displayed along with the computer data on a cathode-ray tube. By moving and zooming the camera under operator control, close-ups of critical map locations could be presented. The problem was indexing the computer data with the map so the computer-generated locations would appear at the correct map locations. One solution was to draw index points on the map and require the operator to place a cursor over each index point (using a mouse or trackball). The universal-transverse-Mercator coordinates (or latitude and longitude) were then associated with each index point, and the computer

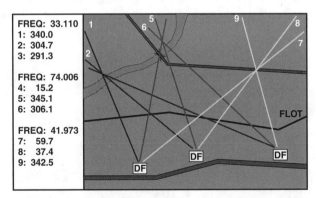

Figure 5.22 In early computerized systems, measured emitter locations were combined with terrain and tactical information in a line-drawing display. Individual lines of bearing could be edited out if they seemed to be caused by multipath, or if they seemed to be erroneous due to multipath or cochannel interference.

could coordinate the map display with the location points it generated. As a point of interest, this is the procedure used to calibrate the touch screen of a hand-held computer after upgrading to a new operating system. This procedure required significant operator tasking and introduced additional (unresolvable) errors if the operator did anything wrong. Also, it was difficult to maintain accurate indexing as the camera moved or zoomed.

5.6.4 Modern Map-Based Displays

Once digital maps became available, it was possible to load a map into a computer and add other information directly to the digital data file. Now the map could be edited in real time to add the tactical situation, the locations of DF receivers, lines of bearing, and any other desired information. Figure 5.23 shows an edited digital-map display. You can see in the figure that the FLOT is located near the top of a steep ridge, and that the DF receivers (one, two, and three) are located on high-terrain features to achieve good lines of sight. You can also see that the three lines of bearing indicate an enemy transmitter location on the eastern edge of Gem Lake. An enemy headquarters is shown (symbolically) to the west of Long Lake.

This display gives the operator a great deal of information to allow the evaluation of the emitter location relative to the tactical situation and terrain features. It can also be zoomed, reoriented, or scanned over to another location without loosing any display accuracy.

This type of display can be easily accessed by several operators and optimized to the needs of each. An analyst or commander can also take a quick look at the raw data to resolve any problem that occurs in higher-level analysis.

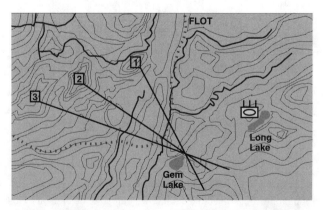

Figure 5.23   A modern ESM operator display combines a digital map with tactical-situation information and overlays the DF-receiver locations and the lines of bearing.

Although the specific digital map used to generate Figure 5.23 is based on a commercial product (used by permission of Wildflower Productions Inc.), military displays use Defense Mapping Agency (DMA) maps. DMA maps contain significant amounts of additional data (terrain surface, etc.) and can be electronically transmitted to a deployed system as required.

6

Search

One of the thorniest problems faced by electronic warfare system designers is detecting the presence of threat signals. Ideally, the receiving part of the EW system would be able to see in all directions at once, at all frequencies, for all modulations and with extremely high sensitivity. While such a receiving system could be designed, its size, complexity, and cost would make it impractical for most applications. Therefore, the practical EW receiving subsystem represents a tradeoff of all of the above-mentioned factors to achieve the best probability of intercept within the imposed size, weight, power, and cost constraints.

6.1 Definitions and Parametric Constraints

Probability of Intercept (POI). This is the probability that the EW system will detect the presence of a particular threat signal between the time it first reaches the EW system's location and the time at which it is too late for the EW system to do its job. Most EW receivers are specified to achieve a probability of intercept of 90–100% for each of the signals in its threat list, when a specified set of signals is present in a specified scenario, within a specified time.

Scan-on-Scan. Scan-on-scan literally refers to the problem of using a scanning receive antenna to detect a signal from a scanning transmit antenna, as shown in Figure 6.1. However, this expression is also used to describe any situation in which a signal must be found in two or more independent dimensions (for example, angle and frequency). The challenge presented by a scan-on-scan situation is that the probability of intercept is reduced, because the periods during which the signal is present vary independently from the periods during which the receiver is able to receive the signal.

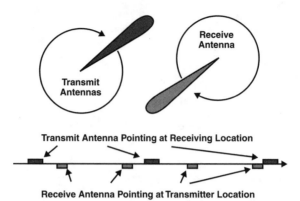

Figure 6.1   In the classical scan-on-scan situation, the transmitting and receiving antennas scan independently from each other. The receiver can only receive the signal when the two antennas are aligned.

6.1.1   Search Dimensions

Now, consider the dimensions in which the EW receiver must find the threat emitter. They are: direction of arrival, frequency, modulation, received signal strength, and time. Table 6.1 shows the impact on the probability of intercept for each of these dimensions.

Direction of Arrival. Particularly for airborne platforms, the direction of arrival is a significant search driver. A maneuvering fighter or attack aircraft can have any orientation, so even signals from threat transmitters that are on the ground can arrive from any direction. Thus, it is normally necessary to consider threats from the complete sphere surrounding the aircraft, which is called "4 π steradian coverage" as shown in Figure 6.2. In aircraft which normally fly with wings more or less level, it is often acceptable to consider only the angular range of 360° in the yaw plane and ± 10° to ± 45° in elevation as shown in Figure 6.3, depending on the mission.

For ship- and ground-mounted EW systems, angular search coverage is typically 360° in azimuth and from the horizon to an elevation of 10–30°, depending on the mission. Although these systems may be required to provide protection from airborne threats, which can fly at any elevation, the relatively small volume at higher elevations means that threat emitters will spend little time at these elevation angles, as shown in Figure 6.4. Another factor is that by the time a platform carrying a threat emitter is observed at a high elevation angle, it is so close that the received signals are at a high enough power level to make their detection hard to avoid.

Table 6.1
POI Impact of Search Dimensions

Search Dimension	Impact on Probability of Intercept
Direction of arrival	Wide angular search volume requires a long search time or dictates a wide beamwidth (i.e., low gain) receiving antenna; both reduce POI.
Frequency	Wide frequency range requires a long search time or dictates a wide bandwidth (i.e., low sensitivity) receiver; both reduce POI.
Modulation	Strong CW or FM signals may interfere with wide bandwidth pulse receivers, reducing POI for pulse signals. Also, CW and FM signals require narrowband receiver types.
Received signal strength	Weak signals dictate narrow beamwidth antennas and/or narrow bandwidth receivers, reducing POI.
Time	Low duty cycle signals can be detected only when they are present in the receiver, extending the required search time of narrow beamwidth antennas and/or narrow bandwidth receivers. When search time exceeds signal up-time, POI is reduced.

Frequency. Radar signals are found throughout the UHF and microwave frequency range. They can also be up into the millimeter wave frequency range. However, detailed knowledge about the potential enemy's

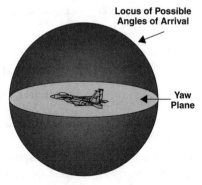

Figure 6.2 For a fighter or attack aircraft, the signal angle of arrival can typically be anywhere on a sphere surrounding the aircraft.

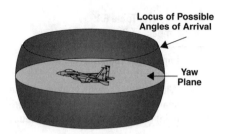

Figure 6.3 For an EW system on an aircraft that normally flies with wings level, the threat angle is typically limited to a band of angular space near the yaw plane.

threat emitters may allow narrowing the search to known frequency ranges of those emitters. Tactical communications signals are found in HF, VHF, and UHF frequency ranges. Each type of emitter can typically be tuned over a wide frequency range, so communications-band receivers must usually search the whole frequency band of interest.

Modulation. In order to receive a signal, it is ordinarily necessary that an EW receiver be configured with the proper demodulator. If signals of widely different modulations are present, this represents another search dimension. The best example is pulsed versus continuous wave (CW) radar signals. CW signals are usually transmitted at significantly lower levels than pulsed signals, and they require different detection approaches.

In searches for communications signals, the initial search is often made with an "energy detection" approach which will simultaneously detect CW, AM, and FM signals. As long as the detection passband is wide enough to contain the modulation, the signal energy is constant. However, single side-band modulation is particularly challenging because the carrier is suppressed, so the received signal strength rises and falls with modulation.

Received Signal Strength. Systems that detect only main beams have strong signals, and can thus use receiver types and search techniques that are

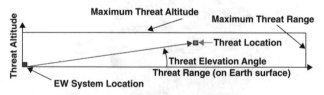

Figure 6.4 For a typical shipboard or ground-based EW system, the maximum threat range is much greater than the maximum threat altitude, so threats are observed at low elevation angles during most of the engagement.

inherently low sensitivity. Systems that detect threats that are not tracking the vehicle carrying the EW system must receive signals from transmit antenna side lobes, which can be 40 dB or more reduced from the signals strength of the main beam.

Time. The time to detect and identify threat emitters is specified (and is short—a very small number of seconds). The EW system must normally find the signal quickly enough to allow time for analysis necessary to identify the signal before it goes away. Since in most cases there are many signals present—any or all of which may be threat signals—time becomes a very critical search dimension.

6.1.2 Parametric Search Tradeoffs

Table 6.2 describes the primary tradeoffs that drive the design of a search approach. In general, the strength of the threat signal at the EW system location and the time available to the EW system to detect it are the two significant factors in the search process. A strong signal will allow the use of wide-beamwidth antennas (which have less gain than narrow-beam antennas) and wide-bandwidth receivers (which have less sensitivity than narrow-bandwidth receivers). Large antenna beamwidths allow the angle-of-arrival dimension to be searched more quickly, and wide-bandwidth receivers allow the frequency dimension to be searched more quickly. Both the angle and the frequency search must be completed within the time allowed for detection of the signal.

Table 6.2
Tradeoffs Between Search Dimensions

Search Dimension	Tradeoff Against	Mechanism
Angle of arrival	Sensitivity	Antenna gain is inversely related to beamwidth
Frequency	Sensitivity	Receiver sensitivity is inversely related to bandwidth
Signal strength	Angle of arrival	Strong signals may allow the use of wide beamwidth antennas
	Time	Receiving system may be able to see threat antenna sidelobes

6.2 Narrowband Frequency Search Strategies

Before covering the sophisticated search approaches made possible by the application of various types of wide-bandwidth receivers, it is instructive to consider what is involved in detecting the presence of a signal in a frequency range that is significantly wider than the bandwidth of a single, independent receiver. We will consider the basic narrowband receiver search strategies for communications and radar signals.

6.2.1 Problem Definition

As shown in Figure 6.5, we will assume that the signal is located in the frequency range F_R (kHz or MHz) and occupies a frequency spectrum of F_M (Hz, kHz, or MHz). (We take this to mean that this amount of spectrum must be within the receiver bandwidth in order for us to detect the presence of the signal.) The search receiver bandwith has units of Hz, kHz, or MHz. The signal is present for a message (or signal) duration of P (seconds or milliseconds). Ordinarily, the search function is time-limited either by the time the signal is expected to be present or by the timing of countermeasures against a lethal threat related to the signal. For communications signals, the requirement is usually to detect the presence of the signal before the message ends, or sufficiently before the end of the message to allow time for analysis,

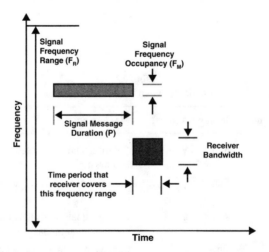

Figure 6.5   The search problem can be visualized as a time versus frequency space in which the receiver bandpass and step dwell time and the target signal band occupancy and message duration are displayed.

location, or effective jamming. For radar signals, the search function must ordinarily find the signal within a fixed time period (typically a small part of a second) to allow for identification and reporting of lethal threats and cueing of countermeasures within a fixed time period after they first illuminate the EW-protected platform (typically some very small number of seconds).

In general, the tuning rate of the searching receiver (the amount of frequency spectrum searched per unit time) must be no greater than one bandwidth in a time equal to the inverse of the bandwidth. For example, if the searching receiver has a 1-MHz bandwidth, it can sweep no faster than 1 MHz per microsecond. In modern, digitally tuned receivers, this translates to dwelling on each tuning step for a period equal to the inverse of the bandwidth. This is often described as "searching at a one-over-the-bandwidth rate." (Note that restrictions on control and processing speed in some receiving systems can further restrict the search rate.)

There are two more restrictions on the search approach: one is that the receiver bandwidth must be wide enough to accept the signal being detected, and the second is that the receiver must have adequate sensitivity to receive the signal with adequate quality. Those of you who are experienced in search applications will note that any of these three restrictions can be mitigated by clever processing if we know something about the signals of interest. This is particularly true when we have an intercept situation in which the target signals will be received with significant intercept margin. However, we will deal with those issues in later sections . . . just remember that nobody, no matter how clever his or her processing, gets to break the laws of physics.

6.2.2 Sensitivity

Naturally, the receiver sensitivity must be great enough to receive the signal you want to detect—that is, received signal strength must be greater than the sensitivity. Remember that sensitivity is defined as the minimum received signal level that will allow the receiver to produce an acceptable output. As explained in Chapter 4, the three components of sensitivity are noise figure (NF), required signal-to-noise ratio (SNR), and the thermal noise level (kTB). When NF and SNR (both in dB) are added to kTB (usually in dBm), the sum is a signal level that is equal to the receiver's sensitivity. The NF is determined by the receiver configuration and the quality of the components. The required SNR depends on the signal modulation and the nature of the information carried. kTB varies primarily with the receiver bandwidth.

$$kTB \text{ (in dBm)} = -114 \text{ dBm} + 10 \log_{10}(BW/1\text{MHz})$$

This means that the optimum search bandwidth is a tradeoff between the sensitivity (i.e., more bandwidth equals less sensitivity) and the speed at which a receiver can be tuned (i.e., more bandwidth equals faster tuning).

The sensitivity required to receive the signal depends upon the intercept geometry. (Received signal strength is described in Chapter 2.)

6.2.3 Communications Signals Search

Since the process of searching for communications signals is simpler in some ways than searching for pulsed signals, we will consider them first. The assumption is made that there is adequate sensitivity to receive the communications signal in the existing intercept geometry.

As shown in Figure 6.6, the basic search strategy is to search as much of the frequency range as possible while the signal is up, using the maximum search rate with as wide a bandwidth as is practical. The main restriction on the bandwidth considered here is its impact on sensitivity, but depending on the signal environment and the signal processing applied, the bandwidth may also be limited by the effect of interfering signals.

6.2.4 Radar Signal Search

Radar signals present additional search challenges, two of which will be considered here. They can be pulsed (whereas communications signals use

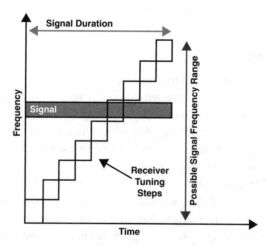

Figure 6.6   In searching for a communications signal, the receiver bandwidth and tuning step duration should allow the whole possible signal frequency range to be covered during the minimum expected signal duration if possible.

continuous modulations) and they have narrow-beam antennas that can be scanned past the receiver location whereas communications signals usually use omnidirectional antennas or fixed wide-beam antennas.

First, consider the intercept impact of pulse signals as shown in Figure 6.7. Since pulsed signals are often high power, there are some other techniques that may be more productive than narrowband search, but there are some circumstances in which it is the only acceptable technique. The signal is present only during the pulse duration (PD), and pulses occur only once per pulse repetition interval. Therefore, the narrowband search receiver must either wait a full pulse repetition interval (PRI) at each tuning step or tune rapidly enough to cover many steps during the PD (and repeat that tuning pattern for a full PRI). If the receiver bandwidth is 10 MHz and the PD is 1 μsec, the receiver can cover only 100 MHz during the PD. Radar signal search bands are typically several GHz wide, so this does not seem to be a very promising search strategy. The alternative of dwelling one full PRI at each tuning step will be slow, but can be improved if a wider receiver bandwidth can be used.

Now consider the effect of a narrow beamwidth scanning transmitter antenna as shown in Figure 6.8. The figure shows the time versus received signal power at the receiver's location. The received power at the time that the transmit antenna is centered on the receiver will be determined by the received power formula shown above. If the receiver sensitivity is adequate to receive a signal 3 dB weaker than this maximum power, the signal can be considered to be present from time *B* to time *C*. If the sensitivity is adequate to receive a 10-dB weaker signal, the signal will be present from time *A* to time *D*. However, if the receiver sensitivity is adequate to receive the sidelobes of the transmitter antenna (see Chapter 3), the signal can be considered to be present 100% of the time.

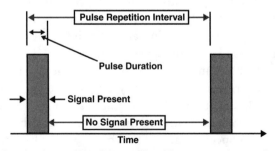

Figure 6.7 During the time a pulsed target signal is present, the receiver only receives energy during pulses, so it must dwell on a frequency for a pulse interval to assure that the signal is observed.

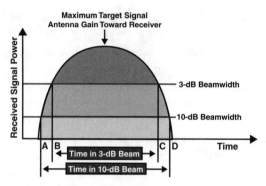

Figure 6.8   As the antenna from a target signal sweeps past the receiver location, its received power varies as a function of time.

6.2.5   Generalities About Narrowband Search

Dealing with finding a signal with a narrowband receiver is a good way to consider the parameters of the search problem faced by EW and reconnaissance receivers. It is not, however, necessarily the best way to conduct a signal search in the real world. After discussing the nature of the signal environment, we will be better equipped to discuss the ways that various types of receivers can be combined to optimize the search in a given environment.

6.3 The Signal Environment

To repeat the often repeated generality, the signal environment is extremely dense and its density is increasing. Like most generalities, this one is usually true but does not tell the whole story. The signal environment in which an EW or reconnaissance system must do its job is a function of the location of the system, its altitude, its sensitivity, and the specific frequency range it covers. Further, the impact of the environment is strongly affected by the nature of the signals the receiver must find and what information it must extract from those signals to identify the signals of interest.

The signal environment is defined as all of the signals that reach the antenna of a receiver within the frequency range covered by that receiver. The environment includes not only the threat signals that the receiver intends to receive, but also signals generated by friendly forces and those generated by neutral forces and noncombatants. There may be more friendly and neutral signals in the environment than threat signals, but the receiving system must deal with all of the signals reaching its antenna in order to eliminate the signals which are not of interest and identify threats.

6.3.1 Signals of Interest

The types of signals received by EW and reconnaisance systems are generally classified as pulsed or continuous wave signals. In this case, "CW signals" include all of the signals that have continuous waveforms (unmodulated RF carrier, amplitude modulation, frequency modulation, etc.). Pulse Doppler radar signals are pulsed but have such a high duty cycle that they must sometimes be handled like CW signals in the search process. In order to search for any of these signals, the receiver must, of course, have adequate bandwidth to receive enough of the signal to observe whatever parameters must be measured. For some kinds of signals, the bandwidth required to detect signal presence is significantly less than that required to recover signal modulation.

6.3.2 Altitude and Sensitivity

As shown in Figure 6.9, the number of signals that a receiver must consider increases directly with altitude and sensitivity.

For signals at VHF and higher frequencies, which can be considered to be restricted to line-of-sight transmission, only those signals above the radio horizon will be in the signal environment. The radio horizon is the Earth surface distance from the receiver to the most distant transmitter for which line-of-site radio propagation can occur. This is primarily a function of the curvature of the Earth, and is extended beyond the optical horizon (an

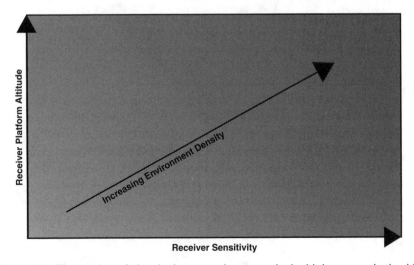

Figure 6.9 The number of signals that a receiver must deal with increases both with increasing platform altitude and increasing receiver sensitivity.

average of about 33%) by atmospheric refraction. The usual way to determine radio horizon is to solve the triangles shown in Figure 6.10. The radius of the Earth in this diagram is 1.33 times the true Earth radius to account for the refraction factor (called the "4/3 Earth" factor). The line-of-sight distance between a transmitter and a receiver can be found from the formula:

$$D = 4.11 \times \left[\sqrt{H_T} + \sqrt{H_R} \right]$$

where D = the transmitter to receiver distance in kilometers; H_T = the transmitter height in meters; and H_R = the receiver height in meters.

Thus the radio horizon has a relative definition—depending on the altitude of both the receiver and any transmitters present. All else being equal, you would expect the number of emitters seen by a receiver to be proportional to the Earth surface area which is within its radio horizon range—but of course the emitter density also depends on what is happening within that range.

For example, an antenna on the periscope of a submarine will receive only signals from the few transmitters expected to be located within a very

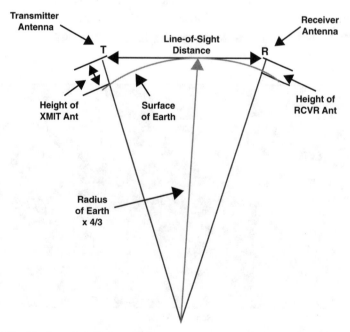

Figure 6.10 The line-of-sight distance between a transmitter and a receiver is determined by the height of both antennas.

few kilometers. While the submarine may see quite a few signals if it is operating close to a large surface task force or to a land area with much activity, the signal density will still be very low when compared to that seen by an aircraft flying at 50,000 feet. The high-flying aircraft can be expected to see hundreds of signals containing millions of pulses per second.

When the receiver is operating below 30 MHz, the signals have significant "beyond the horizon" propagation modes, so the signal density is not so directly a function of altitude. VHF and UHF signals can also be received beyond line of sight, but the received signal strength is a function of frequency and the Earth geometry over which they are transmitted. The higher the frequency and the greater the non-line-of-sight angle, the greater the attenuation. For all practical purposes, microwave signals can be considered to be limited to the radio horizon.

Another element determining signal density is the receiver sensitivity (plus any associated antenna gain). As discussed in detail in Chapter 2, received signal strength decreases in proportion to the square of the distance between the transmitter and the receiver. Receiver sensitivity is defined as the weakest signal from which a receiver can recover the required information—and most EW receivers include some kind of thresholding mechanism so that signals below their sensitivity level need not be considered. Thus, receivers with low sensitivity and those using low gain antennas deal with far fewer signals than high-sensitivity receivers or those which benefit from high antenna gain. This simplifies the search problem by reducing the number of signals that must be considered by the system in identifying threat emitters.

6.3.3 Information Recovered from Signals

It is a reasonable generality that EW and reconnaissance receiver systems must recover all of the modulation parameters of received signals. For example, if the target signal is from a communications transmitter—even though the system may not be designed to "listen to what the enemy is saying"—it will still be necessary to determine the frequency, the exact type of modulation and some of the modulation characteristics to identify the type of transmitter (and thus the type of military asset with which it is associated). For radar signals, the receiver must normally recover the received signal's frequency, signal strength, pulse parameters, and/or FM or digital modulation in order to identify the type of radar and its operating mode.

One significant difference between ESM and reconnaissance receiver systems is that the ESM system will usually recover only enough information about a received signal to allow it to be identified—while the reconnaissance system will normally make a complete set of parametric measurements.

It should be noted that many ESM systems integrate emitter location with the search process, using a preliminary emitter location measurement as part of signal isolation and identification. In the context of signal search, it should be understood that a signal might be classified as friendly or neutral and thus removed from further search consideration based on the location of the emitter.

6.3.4   Types of Receivers Used in Search

The receiver types used in EW and reconnaissance systems include those shown in Table 6.3. Chapter 4 provides functional descriptions of each of these types of receivers. The table shows only their features appropriate to the search problem.

Crystal video receivers provide continuous coverage of a wide frequency range but have limited sensitivity, detect only amplitude modulation, and can only detect one signal at a time. This makes them ideal for handling high-density pulse signal environments, but the presence of a single CW signal can prevent them from accurately receiving any pulses.

IFM receivers provide a digital frequency measurement in an extremely short time but with limited sensitivity. They measure the frequency of each

Table 6.3
Search Capabilities of Receiver Types

Receiver Type	Sensitivity	Can Recover	Multiple Signals?	Instantaneous Frequency Coverage
Crystal video	Poor	Amplitude modulation	No	Full band
IFM	Poor	Frequency	No	Full band
Bragg cell	Moderate	Frequency & signal strength	Yes	Full band
Compressive	Good	Frequency & signal strength	Yes	Full band
Channelized	Good	Frequency and all modulations	Yes	Full band
Digital	Good	Frequency and all modulations	Yes	Moderate range
Superheterodyne	Good	Frequency and all modulations	No	Narrow range

incoming pulse over a full frequency band, but like the crystal video receiver, they can only handle one signal at a time. If one signal is much stronger than the others, the IFM will measure its frequency, but if two or more are nearly the same signal strength, there is no valid frequency measurement. Again, like the crystal video receiver, the IFM is ideal for high-density pulse environments, but a single CW signal can prevent measurements of any pulses.

Bragg cell receivers can measure the frequency of multiple, simultaneous signals so they are not blocked by a single CW signal. However, they currently have limited dynamic range which makes them unsuitable for most EW applications.

Compressive (or "micro-scan") receivers sweep a wide frequency range very quickly—often within a single pulse width. They measure the frequency and received signal strength of multiple simultaneous signals and have good sensitivity; however, they cannot recover signal modulation.

Channelized receivers simultaneously measure frequency and recover the full modulation for multiple signals as long as they are in different channels. They can also provide good sensitivity, depending on the channel bandwidth. However, the narrower the bandwidth, the more channels are required to cover a given frequency range.

Digital receivers digitize a large frequency segment, which is then filtered and demodulated in software. They can measure frequency and recover the full modulation for multiple simultaneous signals and can provide good sensitivity.

Superheterodyne receivers measure frequency and recover any type of signal modulation. They typically receive only one signal at a time, so they are not affected by multiple, simultaneous signals. They can have good sensitivity, depending on the bandwidth. One important feature of superheterodyne receivers is that they can be designed with almost any bandwidth, providing a tradeoff between frequency coverage and sensitivity.

6.3.5 Search Strategies Using Wideband Receivers

There are three basic search strategies used in EW receivers. One is to dedicate one of several receivers to the search function. The second is to determine the frequency of all signals present with wideband frequency measuring receivers and perform detailed analysis or monitoring with set-on receivers. The third is to perform the necessary signal search and measurements with wideband receivers using notch filters and narrowband auxiliary receivers to solve specific signal environment problems encountered.

The first of these strategies is shown in Figure 6.11. This is a common approach for electronic intelligence and communications electronic support

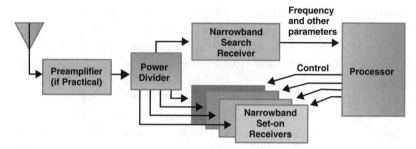

Figure 6.11 In systems that use only narrowband receivers, one receiver is often dedicated to the search function, sweeping at the maximum rate and handing off signals to set-on receivers for complete analysis.

measures systems. The search receiver usually has a wider bandwidth than the set-on receivers and sweeps at the maximum practical rate. It hands off frequency and other quickly measured information on detected signals to a processor which assigns set-on receivers to individual signals to extract all of the necessary detailed information. Note that the antenna output must be power divided for input to each of the receivers because any can be tuned anywhere in the frequency range. Since a power divider reduces the system sensitivity, it is preceded by a low noise preamplifier when practical.

Figure 6.12 shows the second approach. Again, the antenna output must be power divided, and one receiver provides set-on information for multiple narrowband receivers. However, now a wideband frequency measurement receiver is used. The frequency measuring receiver can be an IFM receiver, a compressive receiver, or (if practical) a Bragg cell receiver. Since this receiver can only measure the frequency of the signals present, the processor

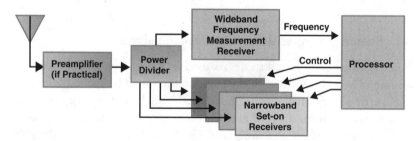

Figure 6.12 A wideband frequency measurement receiver is sometimes used to determine the frequency of all signals present. Then the processor sets narrowband receivers to the optimum frequencies to gather the necessary information from the highest-priority signals.

must assign the set-on receivers based only on frequency. The processor will keep a record of all signals that have been recently found. Typically, it will assign monitor receivers only to new or high-priority signals.

Because some types of frequency measurement receivers have poorer sensitivity than the narrowband set-on receivers, they may not be able to receive some of the signals that could be monitored. This is resolved in two ways. If the received signals are from scanning radars, the less sensitive frequency measurement receiver can detect the signal as its main beam passes the receiving antenna—then the more sensitive monitor receivers may be able to perform their functions by receiving the side lobes of the target emitter. The second mitigating factor is that it often requires less received-signal strength to detect the presence of a signal and measure its RF frequency than is required to get the full signal modulation.

In RWR systems it is typical to have the search function as the primary feature of the system. As shown in Figure 6.13, the typical RWR has a set of wideband receivers (crystal video and/or IFM) to handle the high pulse density. The processor accepts information for each received pulse and performs the necessary signal-identification analysis. Notch filters prevent the blocking of the wideband receivers by CW or high-duty-cycle signals. Narrowband set-on receivers or channelized receivers are used to handle the CW or high-duty-cycle signals and to gather other data that cannot be recovered by the wideband receivers. It is common to have dedicated wideband receivers for each of several frequency bands and a complete set of receivers dedicated to each of several directional antennas to provide pulse-by-pulse direction-of-arrival information.

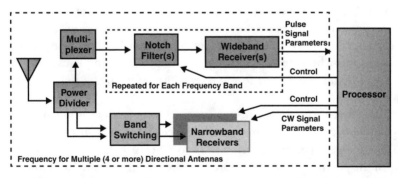

Figure 6.13 For receivers operating in a signal environment dominated by pulsed signals, wideband receivers normally perform the primary search function, protected from CW signals by notch filters. Narrowband receivers handle CW and high-duty-factor signals.

6.3.6 Digital Receivers

Since digital receivers have a great deal of flexibility, they may one day handle the whole search and monitor job. They are restricted by the state of the art in digitization and computer processing (versus size and power requirements)—but the state of the art in these areas is changing almost daily.

6.4 Look-Through

In general, any type of EW receiving system is challenged to detect all of the threat signals present in the brief time available for the search function. There is almost always a wide frequency range to cover, and there are a few signal types that can only be received using narrowband receiver assets. This process is made even more challenging when there is a jammer on the same platform with the receiver or operating in close proximity, because the jammer has the potential to blind the receiver to incoming signals. Consider that EW receiver sensitivities are in the range of -65 to -120 dBm and that jammers typically output hundreds or thousands of watts. The effective radiated power of a 100-W jammer is $+50$ dBm plus antenna gain, so the jammer output can be expected to be 100–150 dB (or more) stronger than the signals for which the receiver is searching.

Whenever possible, the receiver and the associated jammer are isolated operationally—that is, the receiver performs its search function in cooperation with the jammer so that it is searching bands or frequency ranges in which no jamming is momentarily taking place. Where spot jamming and some types of deceptive jamming are used, this operational isolation can allow the receiver to perform a fairly efficient search—if there is some level of isolation to keep the jammer from saturating the receiver's front-end components. Unfortunately, this will seldom solve the whole problem, so other measures must be employed. When broadband jamming is used, a whole band will usually be denied to the receiver unless adequate isolation can be achieved.

The first-choice look-through approach is to achieve as much isolation as possible between the jammer and the receiver. As shown in Figure 6.14, the antenna-gain-pattern isolation is important. Any difference between the gain of the jamming antenna toward the threat being jammed and toward its own receiver antenna reduces the interference. Likewise, any difference between the receiving-antenna gain in the direction of the threats and toward the jammer helps. The antenna patterns shown in the figure are relatively

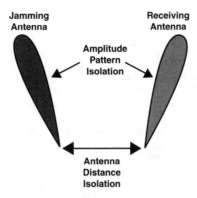

Figure 6.14 The isolation of a receiver from jamming signal power is a function of the distance between the receiving and jamming antennas, the antenna gain pattern isolation, and the polarization isolation.

narrow, but the gain-pattern isolation can just as well be from wide-beam or full-azimuth coverage antennas which are physically blocked from each other (for example, one on the top of an aircraft and the other on the bottom).

The physical separation of the antennas also helps. A formula for spreading loss between two omni-directional antennas is given in Chapter 2. Another form of that equation, but for shorter ranges, is:

$$L = -27.6 + 20 \log_{10}(F) + 20 \log_{10}(D)$$

where L is the spreading loss (in dB); F is the frequency (in MHz); and D is the distance (in meters).

Thus, a jammer operating at 4 GHz and located 10 meters from a receiver would have 64.4 dB of isolation just from the distance between the jamming and receiving antennas.

If the jamming and receiving antennas have different polarization, additional isolation is provided. For example, there is approximately 25 dB of isolation between right- and left-hand circularly polarized antennas. In general, polarization isolation is less than this in wide-frequency-band antennas and can be better than this in very-narrow-band antennas.

Finally, radar-absorptive materials can be used to provide additional isolation, particularly at high microwave frequencies.

If adequate isolation between jamming and receiving antennas cannot be achieved, it will be necessary to provide short look-through periods (as shown in Figure 6.15) during which the receiver can perform its search

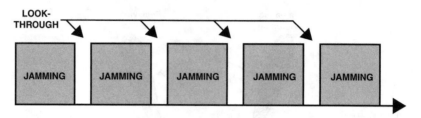

Figure 6.15   If there is not adequate isolation of the receiver from the jammer, it is necessary to interrupt the jamming to allow time for the receiver to perform its
search functions.

functions. The timing and duration of look-through periods are a tradeoff of
jamming effectiveness against probability-of-intercept of threat signals by the
receiver. The look-through periods must be short enough to prevent the
jammed threat radar from receiving adequate unjammed signal to carry out
its mission. On the other hand, the receiver's probability of receiving the
most challenging threat signals during the specified time period will be
reduced by a factor that is strongly related to the percentage of time that the
jammer is transmitting.

7

LPI Signals

7.1 Low-Probability-of-Intercept Signals

Both radars and communication signals are considered low-probability-of-intercept (LPI) signals. LPI radars have some combination of narrow antenna beam, low effective radiated power, and modulation that spreads the radar signal in frequency. LPI communication signals typically depend on the spreading modulation to make them hard to detect and jam. This discussion is focused on LPI communication signals, and more specifically on the frequency spreading modulations that give them advantages over hostile receivers and jammers.

LPI signals are, by design, challenging to the receiving systems attempting to detect them. LPI signals are very broadly defined, including any feature that makes the signal harder to detect or the emitter harder to locate. The simplest LPI feature is emission control—reducing the transmitter power to the minimum level that will allow the threat signal (radar or communication) to provide an adequate signal-to-noise ratio to the related receiver. The lower transmitter power reduces the range at which any particular hostile receiver can detect the transmitted signal. A similar LPI measure is the use of narrow-beam antennas or antennas with suppressed side lobes. Since these antennas emit less off-axis power, the signal is more difficult for a hostile receiver to detect. If the signal duration is reduced, the receiver has less time in which to search for the signal in frequency and/or angle of arrival, thus reducing its probability of intercept.

However, when we think of LPI signals, we most often think of signal modulations, which reduce the signal's detectability. LPI modulations spread the signal's energy in frequency, so that the frequency spectrum of the transmitted signal is orders of magnitude wider than required to carry the signal's information (the information bandwidth). Spreading the signal energy reduces the signal-strength-per-information bandwidth. Since the noise in a receiver is a function of its bandwidth (as described in Chapter 4), the signal-to-noise ratio in any receiver attempting to receive and process the signal in its full bandwidth will be greatly reduced by the signal spreading.

As shown in Figure 7.1, there is a synchronization scheme between the transmitter and the intended receiver that allows the intended receiver to remove the spreading modulation, thereby enabling it to process the received signal in the information bandwidth. Since a hostile receiver is not party to the synchronization scheme, it cannot narrow the signal bandwidth.

There are three ways in which modulation is used to spread the signal in frequency:

- Periodically changing the transmission frequency (frequency hopping);
- Sweeping the signal at a high-rate (chirping);
- Modulating the signal with a high rate digital signal (direct sequence-spectrum spreading).

The challenge that all LPI modulations pose to the search function is that they force an unfavorable tradeoff of sensitivity versus bandwidth. In some cases, the structure of the spreading technique allows some advantage to the receiver, but this requires some level of knowledge about the modula-

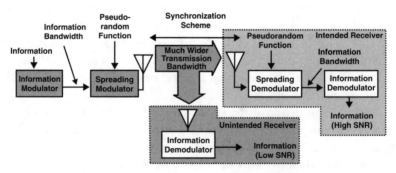

Figure 7.1   Spread-spectrum signals are broadcast with much wider bandwidth than that containing the information they carry. The intended receiver can reduce the bandwidth to the information bandwidth, but unintended receivers cannot.

tion characteristics and can significantly increase the complexity of the receiver and/or its associated processor.

7.1.1 LPI Search Strategies

The basic LPI search techniques always involve optimization of intercept bandwidth and one or more of the following:

- Energy detection with various integration approaches;
- Fast sweep with accumulation and analysis of multiple intercepts;
- Wideband frequency measurement with hand-off to a fast-tuning receiver;
- Digitization and processing using various types of mathematical transforms.

The techniques applied to each type of LPI modulation will be discussed in the section that describes that modulation.

7.2 Frequency-Hopping Signals

Frequency-hopping signals are an extremely important electronic warfare consideration because they are widely used in military systems and because conventional detection, interception, emitter location, and jamming techniques are not effective against them. Although a radar that randomly changes frequency from pulse to pulse can be considered a frequency hopper, we will concentrate on frequency-hopping communication signals.

7.2.1 Frequency Versus Time Characteristic

As shown in Figure 7.2, a frequency-hopping signal remains at a single frequency for a short time, then it "hops" to a different frequency. The hopping frequencies are normally spaced at regular intervals (for example, 25 kHz) and cover a very wide frequency range (for example, 30 to 88 MHz). In this example, there are 2,320 different frequencies which the signal might occupy. The time that the signal remains at one frequency is called the "hop period" or the "hop time." The rate at which it changes frequency is called the "hop rate."

For reasons explained below, frequency-hopping signals carry their information in digital form, so there is a data rate (the bit rate of the

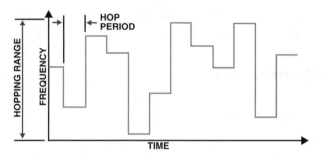

Figure 7.2   A pseudorandom hopping sequence.

information signal) and a hop rate. Signals are described either as "slow hoppers" or "fast hoppers." By definition, a slow hopper is a signal in which the data rate is faster than the hop rate, and a fast hopper has a hop rate faster than the bit rate. However, most people speak of a signal with a hop rate of about 100 hops per second as a slow hopper and a signal with a significantly higher hop rate as a fast hopper.

7.2.2   Frequency-Hopping Transmitter

Figure 7.3 shows a very general block diagram of a frequency-hopping transmitter. First, it generates a signal that carries the information in its modulation. Then, the modulated signal is heterodyned to the transmission frequency by a local oscillator (a fairly fast synthesizer). For each hop, the synthesizer is tuned to a frequency selected by a pseudorandom process. This means that although a hostile listener has no way to predict the next tuning frequency, there is a method by which a cooperating receiver can be

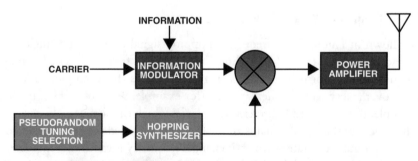

Figure 7.3   A frequency-hopping signal is generated by heterodyning a modulated signal with a local oscillator that is tuned by commands from a pseudorandom tuning frequency-selection circuit.

synchronized to the transmitter. When synchronized, the cooperative receiver tunes with the transmitter, so it can receive the signal almost continuously.

Since a synchronizer takes a small amount of time to settle onto a new frequency (as shown in Figure 7.4), there is a period of time at the beginning of each hop during which no data can be transmitted. This will be a small percentage of the hop period. This settling time is the reason that the information must be transmitted in digital form. The data transmitted during the good portion of the hop can be used to generate a continuous output signal at the receiver—so the human ear won't have to deal with the hopping transitions.

7.2.3 Low Probability of Intercept

The frequency hopper is a low-probability-of-intercept (LPI) signal because the amount of time that it occupies a frequency is too short for an operator to detect a signal's presence. Taking the example above, the signal would be expected to be at any given frequency only 0.04% of the time, so its received power (over time) is significantly reduced, even though its full power is present at a single frequency for the hop period.

7.2.4 How to Detect Hoppers

Actually, frequency-hopping signals (particularly slow hoppers) are easier to detect than some other types of LPI signals, since they have their full power within a single information bandwidth (just like a fixed-frequency receiver) for some period of time (about 10 msec for slow hoppers). A receiver can

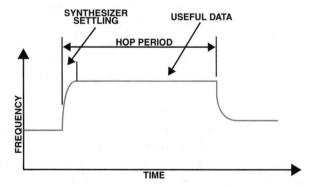

Figure 7.4 The hopping synthesizer will require some period of time to settle before the frequency-hopping transmitter can transmit data.

detect energy in a small part of this time, so it can scan many channels during each hop. Increasing the receiver's bandwidth helps even more, because it covers more of the possible tuning frequencies in each step, and it can step at a higher rate. Another point to remember is that you don't need to cover the full band during a single hop; as long as you catch an occasional hop, you can detect the signal's presence. Naturally, the higher the hop rate, the harder it is to detect the hopping signal.

Wide-frequency-range receiver types that measure frequency (e.g., Bragg Cell, IFM, or compressive) can do a better job of detection, assuming that there is enough received power. Remember that some of the wideband receiver-types have limited sensitivity.

7.2.5 How to Intercept Hoppers

While detecting hoppers is straightforward, intercepting the hopping signal is more challenging. The problem is that you have to detect the hopper and determine its location—then tune to that frequency—before you can start to receive the signal's modulation. Since you don't have any way to predict the frequency of the next hop, you need to repeat the search for every hop. If you are happy with receiving 90% of each hop, you will have to search the whole hopping range in 10% of a hop period (less that settling time). Now you're into some really fancy searching. This probably requires some type of wideband frequency measurement.

7.2.6 How to Locate Hopping Transmitters

There are two basic approaches to performing direction finding (DF) on frequency-hopping signals. One is to sweep the hopping range with a fast-tuning receiver, then perform a quick DF measurement when you find it. This type of DF system typically catches only a moderate percentage of the hops, but it keeps track of the direction of arrival each time it is able to make a measurement. After it gets some number of DF measurements at a single angle of arrival, it reports that there is a frequency-hopping transmitter in that direction.

A second way is to instantaneously cover the whole hopping range, or a major portion of it, in two or more wideband receivers—and to make the DF measurements by processing the outputs of those receivers. If these are digital receivers, the digitized signals are processed using one of several transforms to determine the signal frequency and the direction of arrival. If wideband analog receivers are used, the relative amplitude or phase of the receiver inputs is compared to determine the direction of arrival.

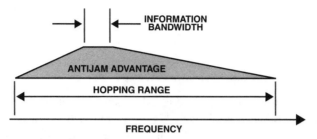

Figure 7.5 In order to continuously jam a frequency-hopping signal with a conventional jammer, the jammer's power must be spread across the whole hopping range. The ratio of the hopping range to the information bandwidth is the antijam advantage.

7.2.7 How to Jam Hoppers

Frequency hoppers are said to have an antijam advantage (see Figure 7.5). This advantage is based on the assumption that the jammer knows only the full hopping range and must spread its jamming power over that full frequency range. In the example used above (2,320 steps of 25 kHz), the frequency-hopping radio can be said to have a jamming advantage of 2,320, which converts to 33.6 dB. This means that it takes 33.6 dB more jammer power to achieve a given jammer to signal ratio against this frequency hopper than would be required if it were a fixed-frequency communication link.

Another disadvantage to this approach is that you will very likely jam every friendly communication link operating within the hopping frequency range. Therefore, two other approaches are used. One is to perform "follower jamming." A follower jammer detects the frequency of each hop and then jams on that frequency. This is an elegant solution, but requires an extremely fast frequency-measurement technique in order to get the jammer on the signal quickly enough to deny the enemy the transmitted information in each hop.

A second approach is to use wideband jamming, but to place the jammer close to the enemy's receiver. This allows effective jamming with minimal jammer power and protects friendly communications.

7.3 Chirp Signals

The second type of LPI signal discussed in this chapter is the "chirped" signal. A chirped radar signal has a frequency modulation during its pulse to allow compression of the received return pulses for improved range resolution. However, when a swept frequency modulation is applied to communication or data signals, its purpose is to prevent detection, intercept or jamming of the signal, or the location of the transmitter.

7.3.1 Frequency Versus Time Characteristic

As shown in Figure 7.6, a chirped signal is rapidly swept across a relatively large frequency range at a relatively high sweep rate. It is not necessary that the sweep waveform be linear as shown in the diagram, but it is important to vulnerability minimization that it be difficult for a hostile receiver to predict when the signal will be at any specific frequency. This can be accomplished by varying the sweep rate (or shape of the tuning curve) in some random way, or to implement a scheme in which the start time of the sweep is pseudo-randomly selected.

7.3.2 Chirped Transmitter

Figure 7.7 shows a very general block diagram of a chirp signal transmitter. First it generates a signal that carries the information in its modulation. Then the modulated signal is heterodyned to the transmission frequency by a local oscillator which is swept at a high rate. The receiver will have a sweeping oscillator synchronized to the transmitter sweep. This oscillator will be used to reconvert the received signal to a fixed frequency. This allows the receiver to process the received signal in the information bandwidth, making the chirp process "transparent" to the receiver. Like the frequency-hopping LPI scheme, the data transmitted can be expected to be digital so that data blocks can be synchronized to the sweeps and then reorganized into a continuous data stream in the receiver.

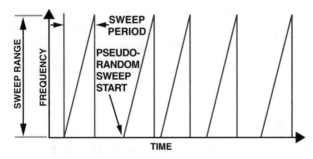

Figure 7.6 A chirped signal is swept across a large frequency range with pseudo-randomly selected start times for its sweep cycles. This precludes a hostile receiver from synchronizing to the chirp sweep.

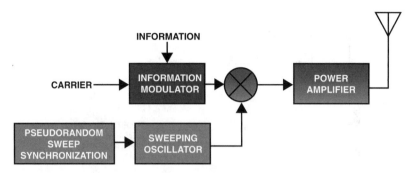

Figure 7.7 A chirped LPI signal is generated by heterodyning a modulated signal with a local oscillator that sweeps across a large frequency range at a high tuning rate. The start time of each sweep is pseudorandomly selected.

7.3.3 Low Probability of Intercept

The LPI qualities of the chirped signal have to do with the way receivers are designed. A receiver typically has a bandwidth approximately equal to the frequency occupancy of the signal it is designed to receive. This provides optimum sensitivity. To maximize transmission efficiency, the signal-modulation bandwidth is approximately equal to the bandwidth of the information it is carrying (or varies by some fixed, reversible factor caused by the modulation).

As discussed in Chapter 6, a signal must remain within a receiver's bandwidth for a time equal to one divided by its bandwidth for the receiver to detect the signal with its full sensitivity. (For example, a 10-kHz bandwidth requires the signal to be present for 1/10,000 Hz, or 100 μsec.) As shown in Figure 7.8, the chirped signal is present in the bandwidth of an information-bandwidth receiver for only a small fraction of the required time.

For example, assume that the information bandwidth is 10 kHz and that the signal is chirped across 10 MHz at a rate of one linear sweep per msec. The swept signal remains within any 10-kHz segment of its 10-MHz sweep range for only 1 μsec—only 1% of the duration required to adequately receive the signal.

7.3.4 How to Detect Chirped Signals

The vulnerability of chirped signals to detection is that their full signal power passes through every frequency within its chirp range. This means that a receiver designed to measure only the frequency of a received signal (without capturing the modulation) may be able to achieve a number of "hits" on a

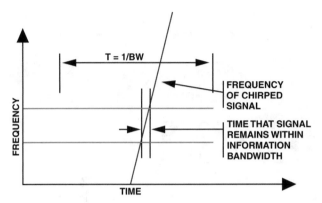

Figure 7.8 An ordinary receiver requires that a signal be within its bandwidth for at least a time equal to the inverse of its bandwidth in order to receive that signal. Because of its high sweep rate, the chirped signal remains within the bandwidth of a receiver optimized to the information signal for much less than this time.

chirped signal. Analysis of this data will show that the signal is chirped and give some level of information about its frequency-scanning characteristics. It is possible to design a "carrier-frequency-only" receiver with greater sensitivity versus instantaneous RF bandwidth than would be required to recover the signal's modulation.

7.3.5 How to Intercept Chirped Signals

To intercept a chirped signal (i.e., recover the information it carries), it is necessary to generate a more-or-less continuous output of the signal's modulation. The obvious way to do this is to provide a sweeping receiver which has a tuning slope the same as that of the chirped transmitter and to somehow synchronize the receiver sweep to the signal sweep.

If the tuning slope can be calculated from a series of carrier-frequency intercepts, generation of the correct receiver-tuning curve is straightforward. Then, if the pseudorandom sweep-synchronization scheme can be solved, the sweep-to-sweep timing can be predicted. Another approach is to digitize the full chirp range and do curve-fitting in software to recover the modulation with some process latency.

Either way, recovering the modulation of a chirp signal with pseudorandom selection of slope or sweep synchronization is technically challenging.

7.3.6 How to Locate Chirped Transmitters

If the chirped signal can be detected, most of the direction-finding techniques discussed in Chapter 8 can be used to locate the transmitter. In general, the implementation of the chosen technique must be such that intermittent reception of the carrier signal is sufficient to measure its angle of arrival. Thus, techniques in which the signal is simultaneously received by two or more antennas seem most suitable.

7.3.7 How to Jam Chirped Signals

Like the frequency hopper, there are two basic approaches to jamming chirped signals. One approach is to predict the frequency-versus-time characteristic of the signal and use a jammer that will input energy to the receiver at the same frequency as the chirped signal it is attempting to receive. This will allow the maximum jammer-to-signal (J/S) ratio to be achieved for any given jammer power and jamming geometry.

The second approach is to cover all or part of the chirp range with a broadband jamming signal that is received by the enemy receiver with adequate power to create adequate J/S ratio in the "dechirped" output. As shown in Figure 7.9, the chirped signal has an antijam advantage equal to the ratio between its information bandwidth and the frequency range over which it is chirped.

Depending on the specifics of the information signal modulation, it may be practical to perform effective partial-band jamming. This jamming technique focuses the jamming power over a fraction of the chirp range that will allow the J/S ratio in the jammed portion to cause a high rate of bit errors in the digital modulation which is carrying the signal's information. The

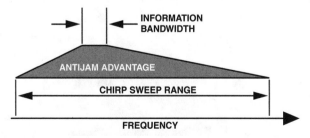

Figure 7.9 Unless a jammer can be swept in synchronization with the signal chirp rate, a jammer's power must be spread across the whole swept range. The ratio of the chirp range to the information bandwidth is the antijam advantage.

fraction of the range which is jammed depends, of course, on the jammer power, the effective radiated power of the chirped transmitter and the relative ranges of the transmitter and the jammer to the jammed receiver. In general, partial-band jamming will cause the maximum disruption of communication for any given jammer power and jamming geometry.

7.4 Direct-Sequence Spread-Spectrum Signals

The final class of spread-spectrum signals discussed in this chapter is direct sequence (DS). This type of signal most closely meets the definition of a spread-spectrum signal since it is literally spread in frequency rather than being rapidly tuned across a wide frequency range. DS has many military and civil applications, since it can protect against both intended and unintended interference and can also provide multiple use of a frequency band.

7.4.1 Frequency Versus Time Characteristic

As shown in Figure 7.10, a DS signal continuously occupies a wide frequency range. Since the DS signal power is distributed over this extended range, the amount of power transmitted within the information bandwidth of the signal (i.e., its bandwidth before it was spread) is reduced by the spreading factor. In Chapter 4 there is a formula for the amount of noise power (kTB) in any given receiver bandwidth. In a typical application, the amount of signal power from a DS spread-spectrum signal will be less than this amount of noise power. Actually, Figure 7.10 is a simplification of the frequency versus time coverage of direct-sequence spread-spectrum signals. The frequency spectrum actually forms a sine(x)/x curve that is very wide compared to the spectrum of the information carried by the signal.

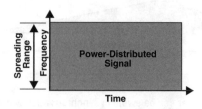

Figure 7.10   The transmitted power of a direct-sequence spread-spectrum signal is spread uniformly across a frequency range that is much larger than that of the basic signal modulation.

7.4.2 Low Probability of Intercept

The low probability of intercept of the DS signal comes from the fact that any noncompatible receiver wide enough to receive the signal will have so much kTB noise that the signal-to-noise ratio of the intercepted signal will be extremely low. This is what is meant by DS signals being "below the noise."

7.4.3 Direct-Sequence Spread-Spectrum Transmitter

Figure 7.11 shows a very general block diagram of a DS spread-spectrum transmitter. First, it generates a signal which carries the information in its modulation. This signal has adequate bandwidth to carry the transmitted information; thus, we say it is an "information-bandwidth" signal. Then, the modulated signal is modulated a second time with a high bit-rate digital signal. One of several phase-modulation schemes is used for this second modulation step. The digital modulating signal has a bit rate (called the "chip rate") that is one or more orders of magnitude higher than the maximum information signal frequency, and it has a pseudorandom bit pattern. The pseudorandom nature of the modulation causes the frequency spectrum of the output signal to spread evenly over a wide frequency range. The power-distribution characteristic varies with the type of phase modulation used, but the effective bandwidth is of the order of magnitude of one divided by the chip rate.

7.4.4 DS Receiver

A receiver designed for DS spread-spectrum signal reception has a spreading demodulator that applies the same pseudorandom signal that was applied by the transmitter (see Figure 7.12). Since the signal is pseudorandom, it has the statistical characteristics of a random signal, but it can be recreated. A

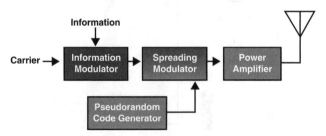

Figure 7.11 A direct-sequence spread-spectrum transmitter modulates the information signal with a pseudorandom digital signal which has a bit rate significantly higher than required to carry the signal's information.

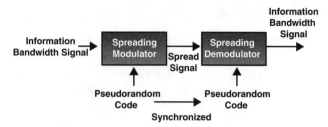

Figure 7.12   When the spread signal is passed through a digital demodulator in a compatible receiver, it is demodulated using the same pseudorandom code that was used in the transmitter, and the receiver code generator is synchronized to that in the transmitter. This restores the information bandwidth signal.

synchronization process allows the receiver code to be brought in phase with the code on the received signal. When this occurs, the received signal is collapsed back down to the information bandwidth, recreating the signal that was input to the spreading modulator in the transmitter.

Since in military applications the spreading code is closely guarded—just as the pseudorandom codes used in encryption are controlled—an enemy trying to intercept the DS signal cannot collapse the signal and must, therefore, deal with the very low power density of the spread transmission.

7.4.5   Despreading Nonspread Signals

A very useful characteristic of the spreading demodulator is that signals which do not contain the correct code are spread by the same factor that the properly coded signal is despread as shown in Figure 7.13. This means that a

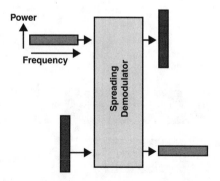

Figure 7.13   The same process that collapses the frequency spectrum of the spread-spectrum signal back to its information bandwidth spreads any nonsynchronized signal by the same factor.

CW signal (i.e., from an unmodulated single-frequency transmitter) that is received by the DS receiver will be spread in frequency and, thus, will have significantly reduced impact on the desired (despread) signal. Since most of the interfering signals encountered in almost any application have narrow bandwidths, a DS link can provide excellent communication in a cluttered environment. This gives the technique significant commercial as well as military applications.

Another reason for the use of DS spread spectrum is to allow multiple use of the same signal spectrum through code division multiple access (CDMA). There are sets of codes which are designed to be mutually "orthogonal." That is, the cross-correlation of any two of the sets is very low. This orthogonality is expressed as a dB ratio: the output of the discriminator is reduced by so many dB if the correct code of the set is not selected.

7.4.6 How to Detect DS Signals

There are two basic ways to detect the presence of a DS spread signal. One is through energy detection with various filtering options. (This is well covered in *Detectability of Spread Spectrum Signals* by Dillard & Dillard, Artech House, 1989.) In general, this requires that the received signal be very strong. The other approach is to take advantage of some characteristic of the transmitted signal. Biphase modulation has a strong second harmonic that is easier to detect. A second characteristic that may be exploitable is the constant chip rate of the spreading modulation. It may be possible to narrow the detection or processing very tightly around spectral lines mathematically related to the chip rate to achieve significant improvement in the detection signal-to-noise ratio.

7.4.7 How to Intercept DS Signals

Like all spread-spectrum signals, DS signals are hard to intercept (i.e., to recover the transmitted information). If the spreading code is known or partially known, it can be applied in sophisticated processing. Otherwise, a wideband digital receiver could capture a segment of the signal from fairly close range and apply various codes in nonreal time to recover the modulation.

7.4.8 How to Locate DS Transmitters

Any of the types of direction-finding approaches with multiple sensors can be used to locate DS transmitters. However, the sensors must be capable of detecting the signal. Then, the received amplitude, phase, or frequency in

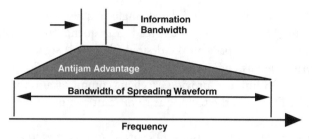

Figure 7.14   In order to jam a spread-spectrum signal, it is necessary to get sufficient jamming energy through the despreading process, which discriminates against
nonsynchronized signals by the ratio of the spreading bandwidth to the information bandwidth.

each sensor can be processed for emitter location. In general, the location of
DS transmitters is fairly easy when strong signals are received, and very complicated for weak signals.

7.4.9   How to Jam DS Signals

As shown in Figure 7.14, the DS spreading provides an antijam advantage
equal to the ratio of the bandwidths. Thus, unless the jammer has significant
information about the spreading modulation, the only practical approach is
to use wideband jamming and to place the jammer close to the enemy's
receiver. This allows effective jamming with minimal jammer power and protects friendly communications for which the receivers are much farther from
the jammer location.

7.5   Some Real-World Considerations

There are some important new techniques and technologies that can be
brought to bear on the challenging tasks of locating and jamming LPI signals.

7.5.1   Spread-Spectrum Signal Frequency Occupancy

Frequency hoppers do not typically occupy a full, contiguous frequency
range, and modern receiving and analysis systems can determine the frequencies that are used. Figure 7.15 (provided courtesy of the company C. Plath
GmbH in Hamburg, Germany) shows typical output from such a system.
This allows emitter location and jamming systems to increase their efficiency
by concentrating on the occupied hop slots.

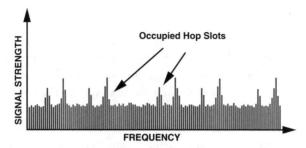

Figure 7.15 The measured frequency spectrum of a frequency-hopping signal shows which channels are being used.

The spectrum occupancy of other types of LPI signals can also be determined with the same types of systems. Figure 7.16 (also supplied by C. Plath) shows the frequency spectrum of a direct sequence spread spectrum signal. This information, of course, reduces the effective "antijam advantage" shown for each of the spread spectrum types we have discussed.

A factor that increases the frequency hopper's detection and antijam advantage is that modern frequency hopping radios can be very selective in their use of hop slots. This allows them to avoid unintentional interference as well as jamming.

Error correction codes can also increase the jamming resistance of hoppers, as discussed in the partial band jamming section below.

While it is appropriate to start by considering the way LPI signals operate without these complications (and many more), it would be easy to be misled by not understanding that the actual implementation is complex and constantly changing. Like most aspects of electronic warfare, this is a highly dynamic contest between communicators (or radars) and the countermeasures.

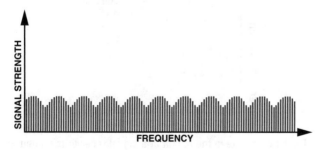

Figure 7.16 The measured frequency spectrum of a direct-sequence spread-spectrum signal shows that its power is not evenly distributed in frequency.

7.5.2   Partial-Band Jamming

This is a jamming technique that optimizes jammer performance against frequency-hopping signals. Frequency-hopping signals carry their information in digital form. When you jam a digital signal, the object is to create enough bit errors to prevent the transfer of useful information from the transmitter to the receiver. The percentage of bit errors that can be tolerated depends on the nature of the information being passed. Some types of information (for example, remote control commands) require extremely low bit-error rates, while voice communication is much more tolerant of errors. Error-correction codes will also make the system less vulnerable to bit errors—and thus to jamming.

As shown in Figure 7.17, there is a nonlinear relationship between the signal-to-noise ratio into a digital receiver and the percentage of bit errors present in the digital output it produces. Communication theory textbooks contain families of these curves—one for each type of modulation technique that is used to carry the digital data. However, all of the curves have the basic shape shown in this typical example. The top of the curve flattens out at about a 50% bit-error rate. This is logical when you think about it—a 50% error rate is as bad as it gets in a digital signal. If the bit-error rate goes above 50%, the output becomes more coherent with the transmitted message. All of the curves reach this 50% point at about 0-dB signal-to-noise ratio (i.e., signal = noise). This means that regardless of the type of modulation used, if the noise level (or the jamming level) is equal to the received-signal level, the percentage of bit errors does not increase when the jamming level is increased.

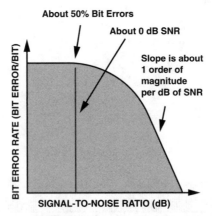

Figure 7.17   The bit-error rate in the output of a digital receiver is related to the in-band signal-to-noise ratio (SNR) into the receiver. Although each type of modulation has a different curve, this one is typical.

Assuming that the transmitter and receiver locations and the transmission effective radiated power (ERP) are known, it is possible to calculate the signal power reaching the receiver. Figure 7.18 shows the communication and jamming geometry. The formula for the signal strength arriving at the receiving antenna (using dB values) is:

$$P_A = \text{ERP} - 32 - 20 \log(d) - 20 \log(F)$$

where P_A = the signal strength arriving at the receiving antenna (in dBm); ERP = the effective radiated power from the transmitting antenna (in dBm); d = the range between the transmitter and the receiver (in km); and F = the transmitted-signal frequency (in MHz).

Based on the above, the equation for the jammer-to-signal ratio (J/S) is defined by the following formula if the receiver antenna beam is omnidirectional:

$$\text{J/S} = \text{ERP}_J - \text{ERP}_T - 20 \log(d_J) + 20 \log(d_T)$$

where J/S = the jammer-to-signal ratio (in dB); ERP_J = the ERP of the jammer (in dBm); ERP_T = the ERP of the transmitter (in dBm); d_J = the range from the jammer to the receiver (in any units); and d_T = the range from the transmitter to the receiver (in the same units).

Ideally, we would transmit enough jammer power, spread across all of the channels to which the transmitter is hopping, to make the power of the jamming signal at the receiving antenna equal to the power of the desired signal (i.e., J/S = 0 dB).

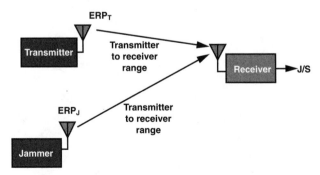

Figure 7.18 Partial-band jamming optimizes the available jamming power by causing the jamming power per channel to equal the received-signal strength from the transmitter over as many channels as possible.

Assuming that our jammer does not have adequate power to provide J/S = 0 dB over the full hopping range, we would narrow the jamming frequency. By concentrating the jamming power in fewer channels until each of the channels jammed has a J/S of 0 dB, we optimize our jamming effectiveness by "partial-band jamming" in this way. If we know enough about the transmitted signal structure to be confident that J/S less than 0 dB provides adequate bit-error rate, a different jammer power distribution may provide optimum results.

7.5.3 Some References for More Study of LPI

The following references are suggested. A spread-spectrum communicatons handbook (1,200 pages) is available from McGraw-Hill. The IEEE has also published a very thorough (1,600 pages) communications handbook that includes much useful information on the subject. *Spread Spectrum Systems with Commercial Applications* by Robert Dixon is available from Wiley; it has excellent information on the details of modulation waveforms and detection and countermeasures techniques. *Principles of Secure Communication Systems* by Don Torrerieri is available from Artech House; it provides excellent coverage of LPI communications, partial-band jamming, and the effects of various countermeasures and counter-countermeasures techniques. In addition to these texts, there is excellent tutorial material along with the data sheets from most of the firms who manufacture receivers and jammers designed to handle LPI signals. The *EW Reference & Source Guide* published by *JED* is an excellent place to start.

8

Emitter Location

Almost all electronic warfare (EW) and signal intelligence (SIGINT) systems require the capability to locate the source of enemy signals. This capability is often called *direction finding* (DF). This is an important concept, because DF specifications are based on required location accuracy and postulated intercept geometry.

8.1 The Role of Emitter Location

EW and SIGINT systems locate signal emitters for several reasons, which are summarized in Table 8.1. In many systems, the information is used in more than one way; the use which requires the greatest location accuracy will determine how the system is designed. The accuracy values in the table are only typical. Requirements for specific applications vary widely. For example, the required location accuracy for determining the electronic order of battle (EOB—the number and types of electronic systems arrayed against you) will depend on the tactical situation, and precision target location accuracy depends on the effective kill radius of the weapons used to attack the located targets.

In many cases, the absolute location accuracy is less important than the resolution provided. *Resolution* is the degree to which the DF system can determine the number of different emitters in its range of operation. Thus, a system collecting emitter location information for EOB needs enough resolution to identify collocated emitters, since this is an important factor in EOB development.

Table 8.1
Implications of Emitter Location Objectives

Objective	Value	Required Accuracy
Electronic order of battle	Locations of emitter types associated with specific weapons and units show enemy strength, deployment, and mission	Medium – ≈ 1 km
Weapon sensor location (self-protection)	Allows focusing of jamming power or maneuver for threat avoidance	Low – general angle and range ≈ 5 km
Weapon sensor location (self-protection)	Allows threat avoidance by other friendly combatants	Medium – ≈ 1 km
Enemy asset location	Allows narrowed reconnaissance search or hand-off to homing devices	Medium – 5 km
Precision target location	Allows direct attack by "dumb bombs" or artillery	High – ≈ 100m
Emitter differentiation	Allows sorting by location for separation of threats for identification processing	Low – general angle and range ≈ 5 km

The increasing use of methods to disguise signals by changing their parameters, such as frequency hopping and jittered pulse repetition frequency, has made the differentiation of emitter location a more important capability for DF systems. Sorting individual pulses or communication signal hops by location may be the only way to determine that they are from the same source, and that may be the only way to collect enough data to identify threat types.

Each of these reasons for measuring emitter location can apply to any type of electromagnetic radiator in any frequency range from "DC to light." Although one approach or one technique may be more suitable for the satisfaction of one of these objectives, most can fulfill any objective if properly specified and designed. The nature of collected signals, the platforms on which the emitter location systems are located, and the predicted tactical situation will usually drive the selection.

8.2 Emitter Location Geometry

Emitter location is performed using one of the five basic approaches described below:

- *Triangulation.* Triangulation techniques locate the emitter at the intersection of two lines from known locations. Figure 8.1 shows this in two dimensions, so the two lines are *azimuths* from which the signal is received at two intercept sites. When the emitter must be located in three dimensions, azimuth and *elevation* are measured from each site. It is highly desirable to have a third intercept site so that the emitter location is defined at the intersection of three lines of bearing. The third bearing provides a "sanity check," since an error in one bearing can cause very large location errors.

- *Angle and Distance.* As shown in Figure 8.2, this technique requires only one intercept site, but it must measure both angle and distance. Most radars locate targets this way, since they are active radiators and directly measure distance, but EW and SIGINT systems must measure distance passively. Single site location (SSL) systems use this approach to determine the distance to HF (between about 3 and 30 MHz) transmitters. Since HF signals are "reflected" from the ionosphere, the distance can be determined by measuring the elevation angle from which the reflected signal is received and the state of the ionosphere (its "height") at the point of reflection. Aircraft radar warning receivers measure received signal power and determine the distance to a radar whose power is already known by calculating the range at which transmission loss would reduce its

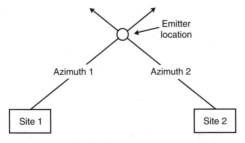

Figure 8.1   Triangulation involves taking direction measurements from more than one source. Where the two-dimensional measurements (azimuths) cross is the likely location of the emitter.

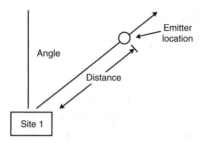

Figure 8.2   In angle and distance techniques, the distance of the emitter from the DF system is derived from the strength of the received signal.

known radiated power to the level received. Both approaches have low accuracy.

- *Multiple Distance Measurements.* The techniques locate the emitter at the intersection of two arcs of known radius. There are two significant problems with practical distance measuring location approaches for EW and SIGINT applications. First, as shown in Figure 8.3, the arcs from the two intercept locations intersect at two points, so some technique must be used to resolve this ambiguity. Second (and typically much harder), it is difficult to passively measure the distance to a noncooperative transmitter with adequate accuracy. *Time difference of arrival* emitter location systems (discussed in detail in Section 8.8) provide very accurate location using a variation of this approach.

- *Two Angles and Known Elevation Differential.* When the difference in altitude between the DF system and transmitter is known, the

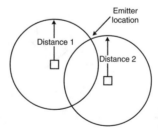

Figure 8.3   Multiple distance measurements locate emitters at the intersection of two arcs. Since the arcs intersect at two points, the system must determine which point is the actual location of the emitter. Also, determining the distance from the edge of each arc to its center is often very difficult.

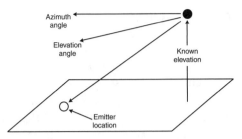

Figure 8.4   If the system knows the difference between the elevation of its host platform and the emitter, measuring the azimuth and elevation angles will determine the location of the transmitter.

transmitter's location can be determined from its azimuth and elevation angles, as shown in Figure 8.4. The best example of this approach is location of ground-based emitters from an aircraft with inertial navigation. The transmit site elevation can be determined from a digital map in the intercept system's computer.

- *Multiple Angle Measurements by a Single Moving Interceptor.* As shown in Figure 8.5, one interceptor can locate a transmitter by taking DF measurements from different locations. However, accurate location requires lines of bearing about 90° apart, requiring the interceptor to travel about 1.4 times the minimum distance to the target—while the transmitter remains on the air and stationary. For distant emitters this time can be excessive, even for airborne interceptors.

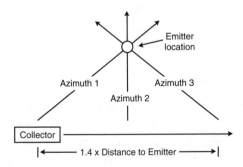

Figure 8.5   Moving interceptors can take several direction measurements, compare them, and determine the location of a fixed emitter.

8.3 Emitter Location Accuracy

The accuracy of emitter location systems is characterized in several different ways, and confusion about the meanings of the terms used causes some of the loudest disagreements between suppliers and users in the EW and SIGINT fields. Accuracy is normally specified in terms of measurement errors. In angle-measuring systems (DF systems) the errors are angular, while in distance-measuring systems the errors are linear. The most common definitions are:

- *Root mean square (RMS) error:* the characterization of the overall effective accuracy of a system over a dimensional range (usually frequency or angle of arrival). It is easiest to talk about the RMS angle-of-arrival error of a DF system, but the same definition can apply to any emitter location approach. Angular RMS error is measured on an instrumented test range where measured angles of arrival can be compared to true angles of arrival. Data is taken at many angles and frequencies. For each data point, the measurement error is squared. The RMS error is the square root of the average of the squared error values. It is common to see the RMS error for all angles at one frequency or the RMS error for all frequencies at one angle.
- *Global RMS error:* the RMS error for a large number of measurements distributed over the whole frequency and angle-of-arrival range.
- *Peak error:* the maximum individual error expected or measured. In practical emitter location systems, there are often a few angle/frequency points at which large errors are measured, particularly during field tests in less-than-optimum site locations. If measured errors are very low at most angles and frequencies, the RMS error can be significantly lower than the peak errors.

8.3.1 Intercept Geometry

In emitter location systems that use triangulation, the intercept geometry is a significant consideration. As shown in Figure 8.6, the location accuracy is a function of both angular measurement error and distance to the emitter being measured. Thus, a distant DF system requires much more angular accuracy to achieve the same location accuracy that a much less accurate system could achieve at closer range.

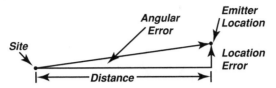

Figure 8.6 The location accuracy generated by a DF system depends on the angular errors and the distance to the emitter.

A second accuracy issue arises from the relative locations of the inter-ceptors to the target transmitter. The term circular error probable (CEP) is often used to describe the location accuracy of an emitter location system. CEP is a bombing and artillery term referring to the radius of an imaginary circle into which half the bombs or shells are expected to fall. In emitter loca-tion, it is sometimes (mis)used to mean a circle that will fit into the space between lines that are ± the RMS error angle from the emitter, as shown in Figure 8.7. The size of the location circle is a function of both the angular error and the distance from the target transmitter to the intercept sites. For the "CEP circle" to be a circle, the two intercept sites must be 90° apart (from the target transmitter's perspective) and at approximately the same distance. When they are less than 90° apart, as shown in Figure 8.8, the unsymmetri-cal area between the lines calls for an ellipse, so the term "elliptical error prob-able" is used to describe location accuracy that is significantly worse in one

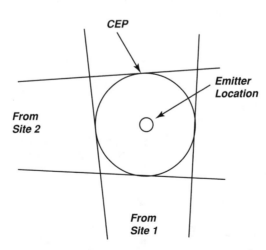

Figure 8.7 "Circular error probable" is a common way to describe the location accuracy provided by angle measurements from two DF sites.

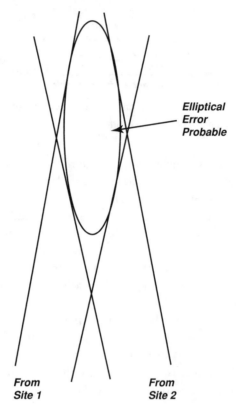

Figure 8.8   "Elliptical error probable" is a common way to describe the nonsymmetrical nature of location accuracy when two DF sites have nonoptimum intercept geometry.

dimension than in the other. This same lack of symmetry will occur if the sites are more than 90° apart or if one site is significantly closer to the target. The term CEP is also applied to these less desirable intercept geometries. It is normally defined as the vector sum of the semi-minor and semi-major axes of the error ellipse, and it is corrected so that there is a 50% probability that the measured emitter location will fall within the CEP radius from the true emitter location.

8.3.2   Location Accuracy Budget

For any type of emitter location system, the location accuracy is a function of both the intrinsic accuracy of the measurement technique and the way the system is mounted and deployed. The location accuracy achieved is most often given in terms of RMS errors in the angle- or distance-measurement data (Figure 8.9). It is defined by the following equation:

$$E_{RMS} = \sqrt{E_L{}^2 + E_I{}^2 + E_M{}^2 + E_R{}^2 + E_S{}^2}$$

where E_L = the location error of the intercept site(s); E_I = the instrumentation error of the system(s); E_M = the installation error of the system(s); E_R = the reference errors; and E_S = the site errors.

This equation provides a fairly good assessment of the location accuracy a system will deliver in operational use if each error source is independent and produces reasonably random errors. Practical fielded accuracy degrades if errors from different sources systematically add or if large peak instrumentation errors are left uncompensated.

- E_L was a significant problem in the early days of emitter location. However, with the availability of low-cost GPS receivers, this error source has become much more manageable.
- E_I is often advertised as the accuracy of a particular emitter location system. It is almost always small compared with installation and site errors.
- E_M can usually be significantly reduced through careful calibration.
- E_R in angle-measurement systems is usually the inaccuracy of the north reference against which azimuth is measured. On small platforms lacking inertial navigation equipment, this can become the

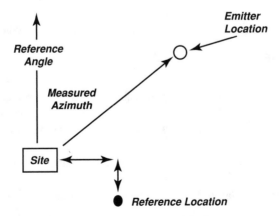

Figure 8.9 Actual location accuracy is a function of both measurement accuracy and reference accuracy.

dominant accuracy-limiting factor for medium- or high-accuracy systems. In very high-accuracy systems, the reference error comes from the reference clock against which time or frequency is measured. The extremely good time/frequency reference available from GPS has diminished this problem in modern systems.

- E_S is usually only a problem in ground-based emitter location systems. Its main causes are multipath reflections from nearby terrain or objects. Site calibration significantly improves the accuracy of fixed sites but is usually impractical for mobile systems.

8.3.3   Emitter Location Techniques

Table 8.2 shows the emitter location techniques commonly used in EW and SIGINT systems, along with typical applications and figures of merit.

Each of these techniques—and the practical problems associated with implementing them in practical, deployed EW and SIGINT systems—will be discussed in later sections.

Table 8.2
Typical Figures of Merit and Applications for Emitter Location Techniques

Technique	Accuracy	Cost	Sensitivity	Speed	Typical Applications
Narrow beam antenna	High	High	High	Low	Reconnaissance and naval ESM
Amplitude comparison	Low	Low	Low	Very high	Airborne radar warning receivers
Watson Watt DF	Medium	Low	Medium	High	Fixed and land-mobile ESM
Interferometer	High	High	High	High	Airborne and land-mobile ESM
Doppler	Medium	Low	Medium	Medium	Fixed and land-mobile ESM
Differential Doppler	Very high	High	High	High	Precision location systems
Time difference of arrival	Very high	High	High	Medium	Precision location systems

8.3.4 Calibration

The accuracy of any type of emitter location system can be improved through calibration. This process involves taking a great deal of data in a controlled situation and determining the measurement errors against the actual location of the test transmitter. For direction-of-arrival type systems, the true angle versus the measured angle is collected. This data is stored in a large computer memory table that is organized by frequency and angle of arrival. For systems that measure other than angle of arrival, appropriate data is collected and stored. Another (and more accurate) way that data can be stored is in terms of the errors in the internal data from which angle of arrival is calculated.

Then, when the system is in operation, collected data is corrected against the calibration table. If the calibration tables are by "type," the same data is used for any emitter location (or direction-finding) system of the type for which data was collected. The calibration tables can also be by "serial number" or "tail number," in which case a unique set of data is collected for each individual system. Calibration by serial number provides greater accuracy but has the disadvantage that it is no longer applicable if anything in the system changes (for example, a failed critical component is replaced).

Calibration is discussed in more detail for the interferometric technique in Section 8.5.

8.4 Amplitude-Based Emitter Location

Of the many ways to locate emitters, techniques that derive location from signal amplitude are usually considered the least accurate. In general, this is true, but these techniques are also (in general) the simplest to implement. Since they can operate successfully on very short-duration signals, amplitude-comparison techniques find wide use in EW systems, although they may be combined with other techniques for emitters requiring precision location. This section will describe three amplitude-based techniques: single directional antenna, Watson-Watt, and multiple antenna amplitude comparison.

8.4.1 Single Directional Antenna

Conceptually, the simplest direction-finding (DF) technique is the use of a single narrow-beam antenna. If only one emitter falls into the antenna beam, and we know the azimuth and elevation pointing angles of the antenna, we know the azimuth and elevation to the emitter. If only the azimuth to the emitter is required, a fan-shaped receiving antenna beam can be used. Figure 8.10 shows the beam pattern of a typical narrow-beam antenna in one dimension. This could be either a parabolic antenna or a phased array. The side and back lobes are normally significantly reduced in gain from the "main beam."

In many shipboard ES (formerly ESM) systems, constantly rotating narrow-beam antennas are used to detect new threat signals at maximum range. The directional antenna approach has many advantages. It isolates individual signals, allowing accurate DF measurement in a dense signal environment (often a problem with other approaches). It provides antenna gain for weak signals, and it can be very accurate. However, this technique has two major problems for some EW applications: it has a significant "scan-on-scan" problem in locating emitters that are present for short periods, and it requires large antennas to provide high accuracy. A related complication is the direct tradeoff of the directional accuracy provided against the scan-on-scan problem (much more about scan-on-scan analysis is discussed in Section 5.3).

To determine the actual location of an emitter using a single directional antenna, it is necessary to get some sort of range measurement. If the radiated power of the emitter is known (often the case for EW threat signals), the range can be estimated from the received power level. Otherwise, range must be determined from some other information (for example, using a known elevation difference, illustrated in Figure 8.4).

8.4.2 Watson-Watt Technique

The technique developed by Sir Robert Watson-Watt in the 1920s is widely used in moderately priced land-mobile DF systems. If three dipole antennas

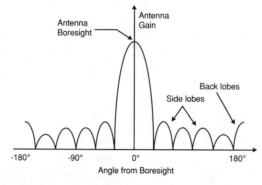

Figure 8.10   The gain of a narrow-beam antenna is much greater near its boresight. Signals at other angles are significantly attenuated.

feed three separate receivers, as shown in Figure 8.11, the coherent sum of the two end antennas (which are separated by approximately 1/4 wavelength) and the center sensing antenna forms a cardiod gain pattern for the array, as shown in Figure 8.12. If the outside two antennas were rotated about the sense antenna, the rotating cardiod pattern would provide direction-of-arrival information on signals at any azimuth. In practical Watson-Watt systems, a number of antennas are arranged as shown in Figure 8.13, and

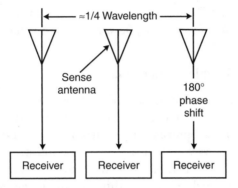

Figure 8.11   The Watson-Watt technique uses two antennas about 1/4 wavelength apart with a central sense antenna.

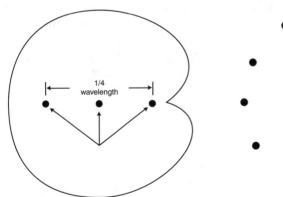

Figure 8.12   The three antennas in a basic Watson-Watt array form a cardiod gain pattern.

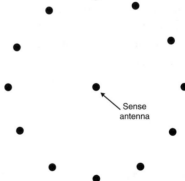

Figure 8.13   Opposite elements in a circular dipole array can be sequentially switched into the Watson-Watt receiver to simulate rotation.

opposite pairs of the outside antennas are sequentially switched into the two appropriate receivers to simulate rotation. The greater the number of antennas, the higher the DF accuracy. But with proper calibration, as few as four antennas can provide acceptable results.

In a further simplification, the function of the central sensing antenna can be provided by the sum of all outside antennas. This allows the Watson-Watt principle to be employed using a simple array of four vertical dipoles, symmetrically placed around an antenna mast. As we will see in later sections in this chapter, the same type of antenna array can be used for several different DF techniques. (The way the antennas are switched into the system and the way the data are processed are, however, quite different.)

8.4.3 Multiple Directional Antennas

Although adaptable to any type of EW system, the multiple directional antenna DF approach is most commonly used in radar warning receiver (RWR) systems. This approach is typically implemented using four or more antennas that have very wide frequency response and stable gain versus angle-from-boresight characteristics. High "front-to-back ratio" (i.e., the ability to ignore signals which are not within 90° of the antenna boresight) is also highly desirable. Ideally, the power gain decreases linearly (in dB) with angle (Figure 8.14). Cavity-backed spiral antennas with gain characteristics close to this ideal (and excellent rejection of signals beyond 90° from the antenna boresight) are used in most modern RWRs.

To understand how this technique determines emitter location, we consider two cavity-backed spiral antennas mounted 90° to each other in the plane of a signal arriving at an included azimuth, as in Figure 8.15. The gain patterns of the two antennas are shown in polar coordinates. Each antenna output is passed to a receiver that measures received power. We note from the

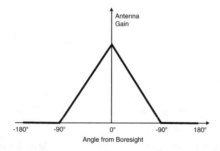

Figure 8.14 The power gain of an ideal amplitude comparison DF antenna varies linearly with angle from antenna boresight out to 90°.

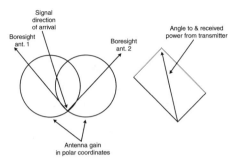

Figure 8.15 The polar representation of two linear gain antennas oriented 90° to each other shows how emitter location is determined from multiple antenna amplitude comparison.

figure that the power received by *antenna 1* is significantly greater than the power received by *antenna 2,* because the path of the arriving signal is closer to the boresight of *antenna 1.* Now we consider the vector diagram. The vector sum of the signals received by the two antennas points toward the transmitter, and its length is proportional to the received power. If the direction of arrival is between the boresights of the two antennas (and they are 90° apart), the direction of arrival and received signal power can be easily calculated from P1 (the power received by antenna 1) and P2 (the power received by antenna 2) as follows:

$$\text{Angle of arrival (relative to antenna 1)} = \text{arc tan (P2/P1)}$$

$$\text{Received signal power} = \sqrt{P1^2 + P2^2}$$

Since the radiated power of EW threat signals is typically known, the received power allows calculation of the approximate range to the emitter, determining the full emitter location.

By placing four such antennas symmetrically around an aircraft as shown in Figure 8.16, 360° azimuthal coverage is achieved. The high front-to-back ratio of the antennas means that only two antennas will receive significant power from any single transmitter, unless the transmitter is located very near one antenna boresight. The irregular shape of the airframe will distort the gain pattern of each antenna, reducing the DF accuracy achieved unless the resultant errors are removed through system calibration. However, achieving high accuracy in either range or azimuth with this type of system requires an extremely complex calibration scheme—so one of the other DF techniques is typically used when more than 5–10° of DF accuracy is required.

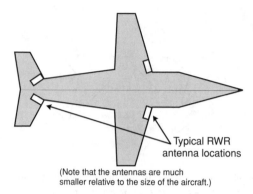

Typical RWR
antenna locations

(Note that the antennas are much
smaller relative to the size of the aircraft.)

Figure 8.16   Four cavity-backed spiral antennas placed symmetrically around an aircraft
allow instantaneous 360° emitter location.

8.5   Interferometer Direction Finding

Interferometry is the most commonly used technique for high-accuracy location of emitters operating at frequencies from just above DC to well above
light. Interferometric systems usually determine emitter location by measuring the angle of arrival (AOA) at two or more DF sites. Systems using this
technique typically provide on the order of 1° RMS angular measurement
accuracy. Interferometer DF finds use in a wide range of electronic warfare
systems, but is most common in radar and communications ES systems.

8.5.1   Basic Configuration

The basic configuration of an interferometer is shown in Figure 8.17. The
key elements are two well-matched antennas—in firmly fixed locations *relative to each other*—feeding two well-matched receivers. An intermediate frequency (IF) output from each receiver is passed to a phase comparator, which
measures the relative phase of the two signals. This relative phase angle is
passed to a processor, which calculates the AOA relative to the orientation of
the two antennas (called the baseline). In most systems, the processor also
accepts information about the orientation of the baseline (relative to true
north or true local horizontal) to determine the true azimuth or elevation
angle to the emitter.

   The biggest challenge in building an interferometer DF system is
to keep the electrical paths through the two antennas and receivers as equal
in length as possible, because the accuracy of AOA measurement depends

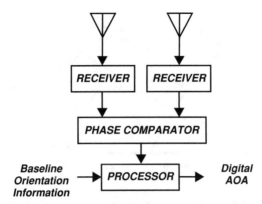

Figure 8.17 The basic interferometer system compares the phase of a signal as received by two matched antennas feeding two matched receivers to determine the signal's angle of arrival.

on the accuracy with which the phase difference between the two receiver outputs can be measured. This requires cables of *exactly* the same length and *exactly* the same phase response through the antennas, receivers and any pre-amplifiers and switches, all the way to the phase comparator—for all signal strengths and at all temperatures. This is such a difficult task that most deployed interferometer systems use some type of real-time calibration scheme to correct phase mismatches. The exceptions are systems in which the antennas and all critical components are built into the same (not too large) box. As you will see in a later section, there are also a number of clever schemes for keeping the critical part of the circuitry close to the antennas in systems where the receivers must be remote from the antennas.

8.5.2 Interferometric Triangle

The interferometric triangle shown in Figure 8.18 describes the way an interferometer DF system determines signal AOA from the relative phase of the signal at the two antennas that form the baseline. The "baseline" is a line connecting the electrical centers of the two antennas, which are rigidly attached to each other. The length of the baseline is *B*, and the AOA of the signal at the baseline is usually referenced to the perpendicular at the center of the baseline. The trick is to measure the value, *d*. Once *d* is known, the AOA is calculated from:

$$AOA = \arcsin (d/B)$$

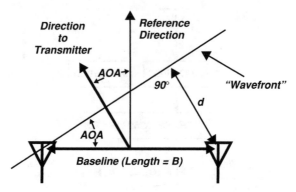

Figure 8.18 The interferometer determines signal angle of arrival relative to its baseline through the interferometric triangle.

An understanding of the interferometric principle is helped by considering an imaginary item called the "wavefront." Electromagnetic waves propagate radially from a transmitting antenna—basic electronics texts often make an analogy to the rings that radiate from a stone dropped into a pond. The "wavefront" is any fixed phase point in the radiating wave as it moves away from the transmitter.

Figure 8.18 depicts the wave front as a straight line because it is a very small segment of a very large circle. Any fixed receiving antenna will observe the propagated wave as a sinusoidally varying signal that moves past it at approximately the speed of light. As shown in Figure 8.19, the phase of the received signal varies by 360° as one full wavelength passes the receiving antenna. By measuring the frequency of the received signal, we can determine the wavelength from:

$$\lambda = c/f$$

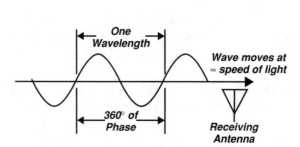

Figure 8.19 The signal propagates at approximately the speed of light, and its phase changes 360° as one wavelength passes a receiving antenna.

where λ = the wavelength (in meters); f = the frequency (in Hz); and c = the speed of light (3×10^8 m/sec).

Then, d is determined by the formula:

$$d = (\phi \times c)/(360 \times f)$$

where ϕ = the relative phase (in degrees) of the signal arriving at the two antennas.

8.5.3 System Configuration

In most practical DF systems, one baseline is not enough; there are ambiguities which are typically resolved by repeating the process with two or more differently oriented baselines, or with baselines of different length. Therefore, the whole system block diagram will be as shown in Figure 8.20. The full set of antennas is called the "antenna array," configured to provide an optimum set of baselines. Each pair of antennas switched into the receivers forms a baseline that is analyzed as described above.

Figure 8.21 shows an array of cavity-backed spiral antennas that might be used to accurately measure the azimuth and elevation angles to a microwave radar transmitter. The horizontally arrayed antennas measure the azimuth, and the vertically arrayed antennas measure the elevation. In each case, the long baseline is switched in to give a very accurate but ambiguous answer, while the short baseline is switched in to resolve the ambiguity.

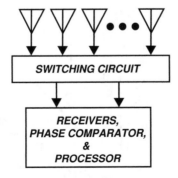

Figure 8.20 The full interferometer DF system includes several antennas which are switched in to the interferometer in pairs to form multiple baselines.

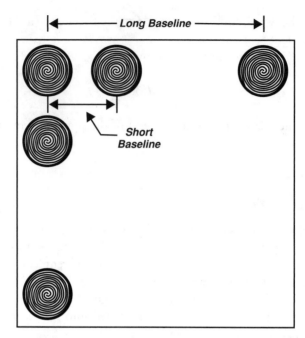

Figure 8.21   Five cavity backed spirals can be formed into an array to feed a high-accuracy, wide-bandwidth, azimuth and elevation interferometer direction-finding system.

Figure 8.22 shows an array of vertical dipole antennas appropriate to a VHF or UHF DF system. In this case, any pair of the four antennas can be selected to form one of six baselines. The diagonal baselines are longer than the side baselines by a factor of 1.414.

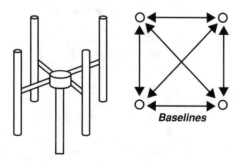

Figure 8.22   Four vertical dipoles can form six baselines for an interferometer direction-finding system.

8.6   Interferometric DF Implementation

To understand the implementation of interferometric DF systems, we must consider inherent ambiguities (and how they are resolved) and the elements controlling accuracy (and how they are improved through calibration).

8.6.1   Mirror Image Ambiguity

First, we must understand that the interferometer simply measures the phase difference between signals arriving at the two antennas that form its baseline. It then converts this information into an angle of arrival. If the two baseline antennas are omnidirectional, a signal arriving from any location on the cone shown in Figure 8.23 would present the same phase difference, causing the interferometer to output the same AOA answer. If we know that the transmitter is located on or near the horizontal plane (a very common situation in ground-based DF systems), this ambiguity reduces to that shown in Figure 8.24. Possible DOAs are now reduced to the two DOAs where the cone passes through the horizontal plane. Without additional information, the interferometer simply cannot tell which of these two answers is correct. If the two antennas are somewhat directional, with high "front to back ratios," resolution is easy, since only one of the answers lies within the area that the two antennas can "see."

Figure 8.25 shows a typical antenna array pattern that might be used in a directional interferometer system. Note that the interferometric principle only works in the area covered by both antennas. However, since this principle operates on the phase difference between the signals received by the antennas, the difference in received signal amplitude caused by a difference in antenna gain in the direction toward the transmitter will have only a secondary effect.

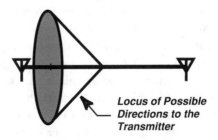

Locus of Possible Directions to the Transmitter

Figure 8.23   The angle of arrival determined by an interferometer defines a cone of possible directions to the transmitter.

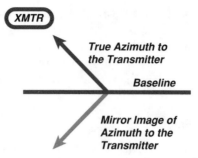

Figure 8.24 The intersection of the cone of Figure 8.23 with the horizontal plane defines two possible azimuths of arrival for the measured signal.

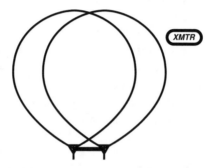

Figure 8.25 If a directional array is used for an interferometer DF system, the target transmitter must fall within the patterns of both antennas forming the baseline.

Since many ground-based DF systems must instantaneously cover 360°, their baseline antennas must be omnidirectional in azimuth (vertical dipoles are very common). These systems resolve the mirror-image ambiguity by making another measurement using a different baseline with another orientation. Figure 8.26 shows this for a 360° ground-based system. The baseline formed by antennas 1 and 3 has one answer in common with the baseline formed by antennas 2 and 4. This is the correct answer.

If an interferometer DF system must deal with signals that are more than a few degrees away from the horizontal plane (almost always the case for systems mounted on aircraft), it is obvious that both azimuth and elevation must be measured to provide an accurate DOA. There is an important exception: if an aircraft is locating emitters known to be on or near the ground and relatively distant, and if the system only considers data received when the wings are close to level, a two-dimensional airborne system can still provide useful data.

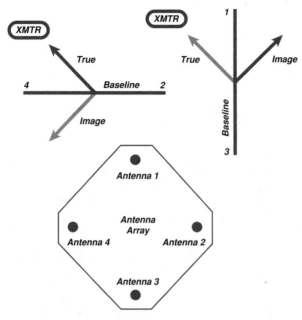

Figure 8.26 Two differently oriented baselines can resolve the mirror image ambiguity in a 360° ground-based interferometer DF system.

8.6.2 Long Baseline Ambiguity

As previously stated, the interferometer determines the AOA by measuring a phase difference between the signals received by the two baseline antennas. At this point, we should consider the relationship between that phase difference and the AOA it represents. This is a function of the AOA, the frequency of the signal and the length of the baseline. The wavelength of the signal (λ) is determined from the speed of light (c) and the frequency (f) by the formula $\lambda = c/f$. Figure 8.27 shows the measured phase difference at the two baseline antennas as a function of the baseline length (in signal wavelengths) and the AOA relative to the system "boresight" (defined as perpendicular to the baseline). The more phase change required to change the AOA (i.e., the steeper the curve on this chart), the more accurate the DF system will be.

Figure 8.27 illustrates two important generalities. First, any interferometer DF system will provide its greatest accuracy for angles near the perpendicular to the baseline (0° in the figure) and its worst accuracy for angles near the ends of the baseline (±90° in the figure). Second, the longer the baseline (relative to the wavelength of the received signal), the greater the accuracy will be.

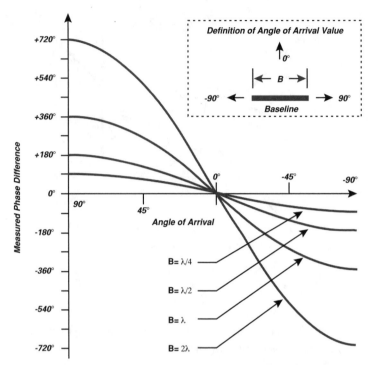

Figure 8.27 The phase difference measured by the two antennas forming an inter-ferometer baseline varies with angle of arrival and with the length of the baseline relative to the wavelength of the received signal.

There is a third, more subtle message in Figure 8.27. Notice that when the baseline gets longer than one-half wavelength, the phase difference changes more than 360° as the AOA moves from +90° to −90°. Since the interferometer has no way of knowing whether or not the signals at the two antennas are in the same cycle of the wave, it will give a very accurate, but ambiguous, answer. This ambiguity is normally resolved by making a separate measurement with a shorter baseline.

8.6.3 Calibration

When an interferometer antenna array is mounted on any type of mast, ground vehicle, or aircraft, the signal received by each antenna will be a combination of the direct wave from the target transmitter and reflections of that wave from everything near the array. Since a reflected wave path is longer than the direct path, each reflection will arrive at the antenna slightly delayed (which will of course give it a different phase). Fortunately, these reflected signals usually have much lower signal strengths than the direct-path signal, but the sum of all of the signals received by each antenna will have a phase that is different from that of the direct-path signal alone. When the relative phase at the two baseline antennas is measured, it will be different than it would have been if only the direct wave signal had been received. This phase difference is called the *phase error,* and it will cause the DF system to derive an incorrect AOA. The difference between the "measured" AOA and the line-of-sight vector from the transmitter to the DF set (called the "true" angle) is called the *angular error.*

To calibrate the system, a set of DF data is taken for every few degrees of AOA and for every few MHz of frequency. Some method is used to determine the true angle (based on the antenna array's known orientation relative to the test transmitter location). Then the angular error is measured for each AOA/frequency combination and stored in a calibration table. Later, when the system is measuring the direction to an unknown transmitter, an appropriate correction factor is calculated by interpolation between points stored in the calibration table.

For interferometer DF systems, the calibration table can contain either the angular error data, as described above, or the phase error for each baseline at each measurement point. In this case, the phase measurements are corrected before the AOA is calculated. Since virtually all interferometer DF systems use several baselines, storing phase data will require significantly more computer memory, but it will produce more accurate results.

8.7 Direction Finding Using the Doppler Principle

Many moderately priced DF systems and some precision emitter location systems use measured changes in the received frequency of signals to determine their direction of arrival. They do this by taking advantage of the Doppler principle.

8.7.1 The Doppler Principle

The Doppler effect changes the received frequency of a signal from its transmitted frequency by an amount proportional to the relative velocity of the transmitter and the receiver. The frequency change can be either positive (when the transmitter and receiver are moving toward each other) or negative (when they are moving apart). In the simplest case, when the movement of one is directly toward the other, the Doppler effect is stated as:

$$\Delta f = (v/c) \times f$$

where Δf = the change in received frequency (called the Doppler shift); v = the speed of the moving element (i.e., the magnitude of the velocity); c = the speed of light (3×10^8 m/sec); and f = the transmitted frequency.

When the transmitter and receiver are not moving directly toward or away from each other, the Doppler shift is proportional to the rate of change in distance between the two, which is:

$$\Delta f = ((V_T \times \cos\Theta_T + V_R \times \cos\Theta_R)/C) \times f$$

where V_T = the speed of the transmitter; Θ_T = the angle between the velocity vector of the transmitter and the direct path between the transmitter and receiver; V_R = the speed of the receiver; and Θ_R = the angle between the velocity vector of the receiver and the direct path between the transmitter and receiver.

If only the transmitter or the receiver is moving, this equation is simplified as the other velocity term goes to zero.

8.7.2 Doppler-Based Direction Finding

The simplest Doppler direction finder is shown in Figure 8.28. Antenna A is fixed, and antenna B rotates around it. Each antenna feeds a receiver, and the received frequency at antenna B is compared to that at antenna A. Figure 8.29 shows the 360° change of antenna B's velocity vector during each rotation. The component of its velocity toward the transmitter is sinusoidal with the positive peak, occurring when it is directly below antenna A in the diagram.

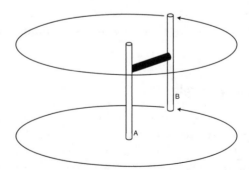

Figure 8.28 A Doppler DF system can be formed by rotating one antenna (B) around a fixed antenna (A).

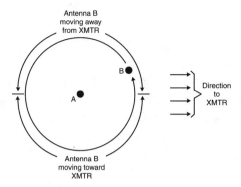

Figure 8.29   The rate of change of the distance from antenna B to the transmitter changes cyclicly as it rotates around the fixed antenna (A).

The difference frequency (antenna B frequency − antenna A frequency) observed for a signal received from any direction varies with time, as shown in Figure 8.30. Antenna B passes between antenna A and the transmitter at the time the Doppler shift goes negative. Since the position of the moving antenna is known to the DF system, the time of this zero crossing is easily converted to the angle of arrival of the signal.

8.7.3   Practical Doppler Direction-Finding Systems

There are obvious mechanical difficulties associated with physically rotating one antenna around another, so most Doppler DF systems use several antennas, circularly arrayed around a central "sense" antenna. The circle of

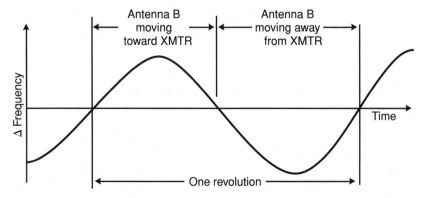

Figure 8.30   The Doppler effect causes the frequency received by antenna B to change sinusoidally relative to the frequency received by antenna A.

antennas you see as you land at most European airports is a Doppler DF array used for passively locating the air-to-ground transmitters on aircraft.

The outside antennas are sequentially switched into a receiver to create the effect of the rotating antenna. In some systems, the sense antenna is eliminated by using the sum of all of the outside antenna outputs as a "reference input." Although a large number of outside antennas will tend to give more accurate answers, this principle will work with as few as three antennas in the "circle." When only a few antennas are used, significant correction factors must be applied to the raw DF data to achieve good DF accuracy.

8.7.4 Differential Doppler

Precision emitter location systems using the Doppler principle are "differential Doppler" systems. They simultaneously measure the Doppler shift at multiple widely spaced receiver locations to determine the transmitter location. This approach can be implemented if either the transmitter or a group of receivers is moving. The movement is, of course, required to generate Doppler shifts. If *both* the transmitter and the receivers have significant velocity, the math gets messy, because each moving element will contribute to the Doppler shift.

At the typical transmitter or receiver speeds encountered in EW applications, the Doppler shift is a very small percentage of the transmitted frequency. The different frequency is easily generated if both signals can be fed directly into a mixer (as in the "rotating antenna" approach). However, comparing the frequency of signals received hundreds of meters (or more) apart requires extremely accurate frequency measurements at each location. Until very recently, this required a local Cesium beam frequency standard, which severely limited differential Doppler applications. However, with the advent of the Global Positioning System (GPS), convenient GPS receivers provide the same frequency standard, making precise frequency measurement relatively easy.

8.7.5 Emitter Location from Two Moving Receivers

The case of two moving receivers locating a fixed transmitter is shown in Figure 8.31. If we know the exact transmission frequency, the measured frequency at each receiver defines the angle between its velocity vector and the transmitter. Thus, if we know the velocity vector (both speed and direction) of each receiver, we can locate the transmitter. This figure assumes that everything is in one plane; location in three dimensions requires three receivers which are not in a straight line.

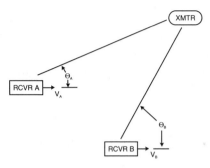

Figure 8.31 Signals from a fixed transmitter will be received by each moving receiver at a frequency that is a function of the speed of that transmitter and the angle between its velocity vector and the direction to the transmitter.

In EW applications, we very rarely know the exact transmitted frequency. The good news is that we can determine some useful information about the transmitter location from just the difference between the two received frequencies. If the speeds of the two receivers are *exactly* equal, the difference frequency will be proportional to the difference of the cosines of the angles between their velocity vectors and the direction to the transmitter. There are an infinite number of transmitter locations that will satisfy this criteria, but they all lie along a curving (precisely definable) line (see Figure 8.32). Since the two receivers will in most cases have at least slightly different speeds, the math gets a little messier, but all of the possible transmitter locations still lie along a curving line that a computer can define. (You could too, but you'd probably die of old age first.)

To determine an unambiguous location, it is necessary to find out where on that line the transmitter lies. The most common way is to make an

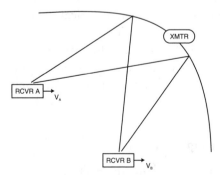

Figure 8.32 The difference in frequencies measured by two moving receivers can be used to calculate a curved line that runs through the transmitter location.

independent frequency difference measurement from another pair of receivers (three receivers can form two independent pairs). The second pair of receivers generates another curve, which will intersect the first at the transmitter location. Location in three dimensions requires three independent receiver pairs.

8.8 Time of Arrival Emitter Location

When precise emitter location is required, the time of arrival (TOA) or time difference of arrival (TDOA) techniques are often the best choices. Both depend on the fact that signals propagate at approximately the speed of light (c), which is approximately 3×10^8 m/sec.

A signal that leaves the transmitter at some defined time will arrive at the receiver at time d/c later, where d is the distance from the transmitter to the receiver. (For example, if d is 30 km, the signal will arrive $30 \div 3 \times 10^8$ sec = 100 μsec after it leaves the transmitter.) Thus, the time of arrival defines the distance. The accuracy with which the distance is defined depends on the accuracy with which the transmission time is known and the time received is measured. (The signal travels about 1 ft per nsec.) GPS receivers output very accurate time references, making precision TOA measurement much easier (logistically) than it was only a few years ago.

If two receivers are placed at known locations, a signal is transmitted at a known time and the arrival time of the signal is accurately measured at each receiving site, then the transmitter's location is defined by the calculated distance from the two receivers. This is true only if the transmitter and receivers are in the same plane (for example, when the transmitter and receivers are all within line of sight and close to the same elevation). In free space the two distances describe a circle (imagine tying two strings to a key and, holding one string in each hand, swinging it in a vertical circle). If the transmitter is known to be on the Earth's surface, its location is, of course, at one of the two points at which this circle intersects the surface. If the receiving antennas have a significant front-to-back ratio, only one of these points applies—otherwise the "mirror image" ambiguity must be resolved by use of two or more TOA baselines.

8.8.1 TOA System Implementation

There are two principal ways in which TOA emitter location systems are implemented, depending on the separation of the receiver locations. If the two receivers forming the baseline are mounted on the same physical structure (for example, in an array or on different parts of the same aircraft), it may be practical to implement the system as shown in Figure 8.33. By carefully matching the antennas, receivers, and cables, the arrival times can be measured in a single processor. If the internal transmission times (*tt*) were exactly the same, the time of arrival at each antenna could be determined by subtracting *tt,* and the difference of arrival times at the processor would exactly equal the arrival-time difference at the antennas.

As a practical matter, manufacturing tolerances and the effects of temperature differences, component aging, and other environmental influences usually create a need for some real-time measurement of the electrical distance from the antenna to the processor for each receiving path. A correction factor can then be applied.

If the receivers are far apart (on different aircraft or ground stations), the implementation shown in Figure 8.34 is required. In this case, precise time measurements are made at each receiver location, and those time values are transmitted to a processor that does the calculations and locates the transmitter.

8.8.2 Time Difference of Arrival

A true TOA approach requires that we know the time at which the signal left the transmitter (i.e., it requires that the signal include some sort of decodable time reference). This is seldom the case in EW applications, but, fortunately, we can determine something about the emitter's location from the difference in the times at which its signal arrives at two receivers. If everything is in one

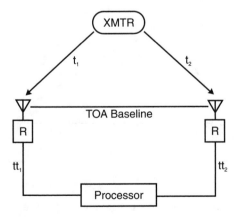

Figure 8.33   If the two receivers are reasonably close, TOA emitter location can be implemented with calibrated cables to the processor.

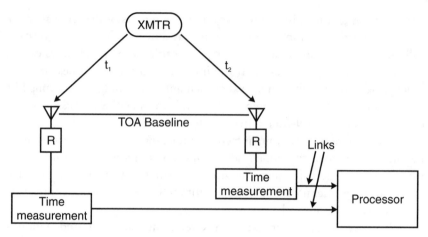

Figure 8.34 TOA emitter location with widely spaced receivers requires precision TOA measurement at each receiver site.

plane, the time difference defines a curving (but mathematically definable) line that will pass through the transmitter (see Figure 8.35). To determine the position of the transmitter along that line, it is necessary to use another TDOA baseline (requiring one more receiver) to generate a curve that will cross the first curve at the transmitter location.

All of the TOA implementation discussion above applies equally well to the TDOA approach, except that one additional receiver is required in each case. That is, three receivers (forming two independent baselines) are

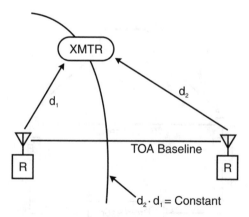

Figure 8.35 The difference in time of arrival at the two receivers defines a line which passes through the transmitter location.

required for emitter location in two dimensions, and three non-coplanar receivers (forming three independent baselines) are required for three dimensions. Because EW applications usually rely on TDOA, the balance of this discussion will focus on TDOA.

8.8.3 Distance Ambiguities

If a signal repeats itself within the time required for it to travel from the farthest possible transmitter location (i.e., the horizon) to a receiver, there will be a distance ambiguity, since a receiver has no way of knowing which repetition it is receiving. Each receiver will be subject to one distance answer per possible repetition, so the number of location ambiguities will be the square of the number of repetitions.

8.8.4 Time of Arrival Comparison

The location of unmodulated CW transmitters by TOA or TDOA is not practical, because they absolutely repeat themselves every RF cycle (causing an infinite number of ambiguities). The repetition time of the modulation is typically much slower, because the modulating waveform is at a significantly lower frequency than the RF. Modulation on information carrying signals is even less likely to repeat because of the nonrepetitive nature of the information.

In order to measure the time of arrival of a signal, we must define an identifiable time reference in the signal's modulation. This requires different approaches for pulse signals and continuously modulated signals.

8.8.5 Pulsed Signals

Pulse signals are designed for ease of time measurement, which is their function in radars. The obvious choice is to simply time the leading edge of the pulse. The difference in arrival time for the leading edge in the two baseline receivers is then the TDOA. In typical EW situations, the leading edges will not be straight or clean, but it is still relatively easy to pick a spot on a pulse to measure.

All of the pulses from a transmitter look alike, and they are repeated at the pulse repetition interval (PRI). Unless there is some type of pulse coding, the TOA distance measurement will only be unambiguous within the distance a pulse travels during one PRI (e.g., 30 km for a signal with a pulse repetition frequency of 10,000 pulses/sec). If a precision TDOA system is used in conjunction with a lower-accuracy location system, the less accurate system may be able to eliminate all of the incorrect locations.

8.8.6 Continuously Modulated Signals

An amplitude-modulated signal looks something like the waveforms in Figure 8.36 when viewed on a fast oscilloscope. Signals 1 and 2 on the figure are short pieces of the same signal, offset in time as they would be if they were received by receivers at different distances from the transmitter. We can see that if *signal 1* is delayed the correct amount, the signals will overlay each other. This means that their "correlation" will be very high.

In a TDOA system, the output of each receiver is digitized. The time tagged digitized characterizations of the signals are sent (by data link) to a processor. In effect, the processor "slides" one signal across the other in time and measures the correlation of the two signals as a function of the amount of delay imposed.

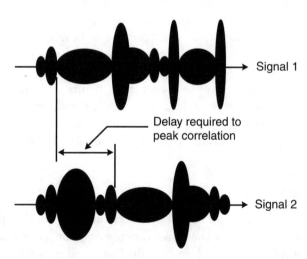

Figure 8.36   Definition of time difference for analog modulated signals is determined by delaying one and measuring correlation.

9

Jamming

The purpose of all jamming is to interfere with the enemy's effective use of the electromagnetic spectrum. Use of the spectrum involves the transmission of information from one point to another. This information can take the form of voice or nonvoice (e.g., video or digital format) communications, command signals to control remotely located assets, data returned from remotely located equipment, or the location and motion of friendly or enemy assets (land, sea, or air).

For many years, jamming has been called electromagnetic counter-measures (ECM), but it is now referred to in most literature as electronic attack (EA). EA also includes the use of high levels of radiated power or directed energy to physically damage enemy assets. Jamming is sometimes called "soft kill" because it temporarily makes an enemy asset ineffective but does not destroy it.

The basic technique of jamming is to place an interfering signal into an enemy receiver along with the desired signal. Jamming becomes effective when the interfering signal in the receiver is strong enough to prevent the enemy from recovering the required information from the desired signal, either because the information content in the desired signal is overwhelmed by the power of the jamming signal or because the combined signals (desired and jamming) have characteristics that prevent a processor from properly extracting or using the desired information. Table 9.1 defines several ways in which the various classes of jamming are differentiated. Subsequent sections will further define many subclasses and specific techniques.

Rule One of Jamming. The most basic concept of jammer application is that you jam the *receiver*, not the *transmitter*. The analysis of a jamming situation is often confusing, and it is easy to make that mistake, so remember

Table 9.1
Types of Jamming

Type of Jamming	Purpose
Communications jamming	Interferes with enemy ability to pass information over a communications link
Radar jamming	Causes radar to fail to acquire target, to stop tracking target, or to output false information
Cover jamming	Reduces the quality of the desired signal so it cannot be properly processed or so that the information it carries cannot be recovered
Deceptive jamming	Causes a radar to improperly process its return signal to indicate an incorrect range or angle to the target
Decoy	Looks more like a target than the target does; causes a guided weapon to attack the decoy rather than its intended target

that to be effective, the jammer must get its signal into the enemy's receiver—through the associated antenna, input filters, and processing gates. This, in turn, depends on the signal strength the jammer transmits in the direction of the receiver and the distance and propagation conditions between the jammer and the receiver.

9.1 Classifications of Jamming

Jamming is typically classified in four ways: by type of signal (communications versus radar); by the way it attacks the jammed receiver (cover versus deception); by jamming geometry (self protection versus stand-off); and by the way it protects a friendly asset (decoy versus classical jammer).

9.1.1 Communications Versus Radar Jamming

Communications jamming (COMJAM) is the jamming of communications signals. This is normally considered the jamming of tactical HF, VHF, and UHF signals using noise-modulated cover jamming, but it can also mean the jamming of point-to-point microwave communications links or command and data links to and from remote assets. As shown in Figure 9.1, the enemy's communications link carries a signal from a transmitter (XMTR) to a

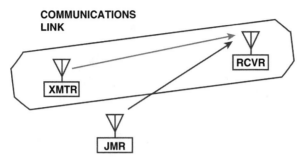

Figure 9.1 Communications jamming interferes with the ability of a receiver to recover information from its desired signal.

receiver (RCVR). The jammer (JMR) also transmits into the receiver's antenna, but it has enough power to overcome the disadvantage of antenna gain (if the receiving antenna has a narrow beam and is pointed at the transmitter) and to be received and output to the receiver's operator or processor with adequate power to reduce the quality of the desired information to an unusable level.

A classical radar has both a transmitter and a receiver, which use the same directional antenna. The radar receiver is designed to optimally receive return signals from objects illuminated by the radar transmitter. Analysis of the return signals allows the radar to determine the location and velocity of some land, sea, or air asset and to track it—for either friendly (e.g., air traffic control) or unfriendly (e.g., attack by guided missiles or guns) purposes. The radar jammer provides either a cover or deceptive signal to prevent the radar from locating or tracking its target (see Figure 9.2).

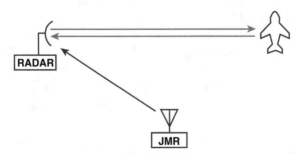

Figure 9.2 Radar jamming, which can be either cover or deceptive jamming, interferes with the ability of radar to recover information about the target from its return signal.

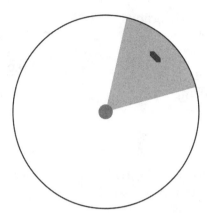

Figure 9.3   Cover jamming hides the radar's return signal from the receiver/processor.

9.1.2   Cover Versus Deceptive Jamming

Cover jamming involves the transmission of high-power signals into an enemy transmitter. The use of noise modulation makes it more difficult for the enemy to know that jamming is taking place. This reduces the enemy's signal-to-noise ratio (SNR) to the point where the desired signal cannot be received with adequate quality. Figure 9.3 shows a radar plan position indicator (PPI) scope screen with a return signal and noise cover jamming strong enough to hide the return. Ideally, the jamming should be strong enough to make it impossible for a trained operator to detect signal presence, but if it is impossible (or impractical) to get that much jamming power into the receiver, it may be sufficient to reduce the SNR enough so that automatic tracking cannot be performed. (Automatic processing usually requires significantly better SNR than that required for a trained operator to detect and manually track a signal.)

Deceptive jamming causes a radar to draw the wrong conclusion from the combination of its desired signal and the jamming signal, as illustrated in Figure 9.4. Normally, this jamming seduces the radar from the target in range, angle, or velocity. With deceptive jamming, the radar gets an apparently valid return signal and "thinks" it is tracking a valid target.

9.1.3   Self-Protection Versus Stand-Off Jamming

Self-protection jamming and stand-off jamming are illustrated in Figure 9.5. Both are normally categorized as radar jamming, but they can be any sort of

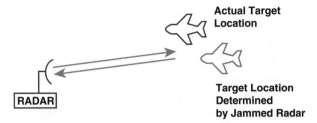

Figure 9.4 Deceptive jamming interferes with the radar's processing to create false information about the target's location or speed.

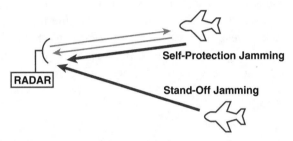

Figure 9.5 Self-protection jamming is provided by jammers on the platform being targeted by a radar. Stand-off jamming allows a high-power jammer on another platform to protect the platform that is being targeted.

jamming that is applied to protect a friendly asset (for example, jamming the communications net used to coordinate an attack). Self-protection jamming originates in a jammer carried on the platform that is being detected or tracked. Stand-off jamming involves a jammer on one platform transmitting jamming signals to protect another platform. Normally, the protected platform is within the lethal range of the threat and the stand-off jammer is beyond the lethal range of that weapon.

9.1.4 Decoys

A decoy is a special kind of jammer designed to look (to an enemy radar) more like a protected platform than the protected platform itself. The difference between decoys and other types of jammers is that a decoy does not interfere with the operation of the radars tracking it, but rather seeks to attract the attention of those radars, causing them to either acquire it and attack it or to transfer the tracking focus.

9.2 Jamming-to-Signal Ratio

The effectiveness of a jammer is calculable only in the context of the enemy receiver that it jams. (Remember, we jam the receiver, not the transmitter.) The most common way to describe that effectiveness is in terms of the ratio of the *effective* jammer power (that is, the jamming signal power that gets into the heart and soul of the receiver) to the signal power (that the receiver really wants to receive). This is called the jamming-to-signal ratio, or the J-to-S ratio, or simply the J/S.

There are many special cases in which this straightforward explanation of J/S must be modified for accuracy—and we will cover the most important of these later—but all are based on the principles explained below. The dB form equations used in this discussion include numerical "fudge factors" (e.g., "32") to take care of various "laws of physics" constants and allow us to input parameters and get answers directly in the most useful units. In this discussion, all distances are in km, all frequencies are in MHz, and radar cross-section (RCS) is always in m².

9.2.1 Received Signal Power

First, let's consider the signal part of the J/S. In the case of a signal transmitted one way from a transmitter to a receiver (as in Figure 9.6), the signal arrives at the receiver input with a power level defined by the equation (all in dB values):

$$S = P_T + G_T - 32 - 20 \log(F) - 20 \log(D_S) + G_R$$

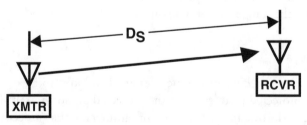

Figure 9.6 The desired signal arrives at the receiver input with its strength determined by the transmitter power, both antenna gains, and a link loss related to its frequency and the link distance.

where P_T = transmitter power (in dBm); G_T = transmit antenna gain (in dB); F = transmission frequency (in MHz); D_S = distance from the transmitter to the receiver (in km); and G_R = receiving antenna gain (in dB).

For the case of a radar signal (as in Figure 9.7), the transmitter and receiver are typically collocated and share the same antenna, so the signal arrives at the receiver with a power level defined by the equation (again, all in dB values) that was derived in Chapter 2:

$$S = P_T + 2G_{T/R} - 103 - 20 \log(F) - 40 \log(D_T) + 10 \log(\sigma)$$

where P_T = transmitter power (in dBm); $G_{T/R}$ = transmit/receive antenna gain (in dB); F = transmission frequency (in MHz); D_T = distance from the radar to its target (in km); and σ = radar cross-section of the radar's target (in m^2).

9.2.2   Received Jamming Power

Jamming signals are, by their nature, one-way transmissions (Figure 9.8). In general, the performance of the jamming signal is the same whether its target is a communications receiver or a radar receiver. Its acceptance by the receiver differs from that of the desired signal in two ways. First, unless the receiver has an omnidirectional antenna, the antenna gain will vary as a function of the azimuth or elevation from which the antenna receives signals. Thus, the jamming and the desired signal will experience different receiving antenna gains (Figure 9.9) unless they arrive from the same direction. Second, jamming signals must often be much wider in frequency than the signals they are jamming because the desired signal's exact frequency cannot be measured or

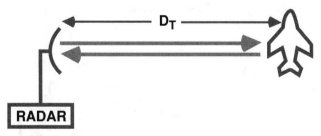

Figure 9.7   A radar signal arrives at the receiver with a signal strength determined by twice its antenna gain, the round trip distance to its target, the signal frequency, and the radar cross-section of the target.

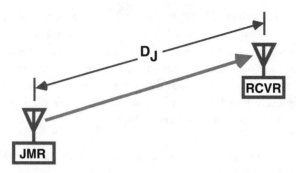

Figure 9.8   The jamming signal arrives at the receiver input with its strength determined by the transmitter power, jammer antenna gain, link loss related to its frequency, and the link distance and the receiving antenna gain in the direction of the jammer.

predicted. In predicting the J/S, it is important to count only the part of the jamming signal power that falls within the receiver's operating bandwidth. With these two understandings, the jamming power arriving at the input to the receiver is defined by the equation (in dB):

$$J = P_J + G_J - 32 - 20 \log(F) - 20 \log(D_J) + G_{RJ}$$

where P_J = jammer transmit power (in dBm) (within the receiver's bandwidth); G_J = jammer antenna gain (in dB); F = transmission frequency (in MHz); D_J = distance from the jammer to the receiver (in km); and G_{RJ} = receiving antenna gain in the direction of the jammer (in dB).

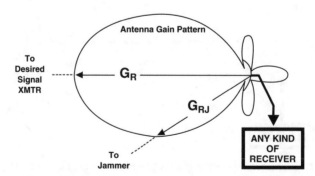

Figure 9.9   If the receiving antenna is not omnidirectional, its gain to the jamming signal will be different (usually less) than its gain to the desired signal.

9.2.3 Jamming-to-Signal Ratio

As shown in Figure 9.10, the J/S is the ratio of the jamming signal strength (within the receiver's bandwidth) to the strength of the desired signal. If dB units are used, the ordinate scale in this figure will be linear. It is, of course, assumed that the receiver bandwidth is ideally sized and tuned to the desired signal. From the above formulas, the development of the J/S formula is straightforward. Since both J and S are expressed in dB, their power ratio is simply the difference between their dB values. For the one-way signal transmission case (applicable mainly to consideration of communications jamming), the J/S in dB is:

$$J/S \text{ (in dB)} = J - S = P_J + G_J - 32 - 20 \log(F) - 20 \log(D_J) + G_{RJ} - [P_T + G_T - 32 - 20 \log(F) - 20 \log(D_S) + G_R] = P_J - P_T + G_J - G_T - 20 \log(D_J) + 20 \log(D_S) + G_{RJ} - G_R$$

Consider, for example, the situation in which the jammer's transmitting power is 100W (+50 dBm), its antenna gain is 10 dB, and its distance from the receiver is 30 km; the desired signal transmitter is 10 km from the

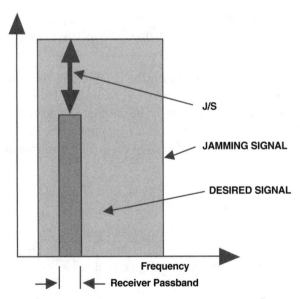

Figure 9.10 The jamming-to-signal ratio is simply the ratio of the power of the two received signals within the frequency passband of the receiver.

receiver, its transmit power is 1W (+30 dBm), its antenna gain is 3 dB, and the receiver's antenna provides 3-dB gain to both the desired and jamming signals. J/S is then calculated to be:

$$J/S = +50 \text{ dBm} - 30 \text{ dBm} + 10 \text{ dB} - 3 \text{ dB} - 20 \log(30)$$
$$+ 20 \log(10) + 3 \text{ dB} - 3 \text{ dB} = 17 \text{ dB}$$

For a jammer operating against a radar, the formula is:

$$J/S \text{ (in dB)} = J - S = P_J + G_J - 32 - 20 \log(F) - 20 \log(D_J) + G_{RJ}$$
$$- [P_T + 2G_{T/R} - 103 - 20 \log(F) - 40 \log(D_T) + 10 \log(\sigma)] = 71$$
$$+ P_J - P_T + G_J - 2G_{T/R} + G_{RJ} - 20 \log(D_J) + 40 \log(D_T) - 10 \log(\sigma)$$

Consider the situation in which the radar's transmit power is 1 kW (+60 dBm), its antenna gain is 30 dB, the range to a 10 m² target is 10 km, the jammer transmits 1 kW into a 20-dB antenna 40 km from the radar, and the jamming signal is received by a 0-dB radar antenna side lobe. J/S is then calculated to be:

$$J/S = 71 + 60 \text{ dBm} - 60 \text{ dBm} + 20 \text{ dB} - 2(30 \text{ dB}) + 0 \text{ dB}$$
$$- 20 \log(40) + 40 \log(10) - 10 \log(10) = 29 \text{ dB}$$

Now consider the case in which the jammer and the target are collocated (e.g., a "self-protection" jammer on an airplane being tracked by the radar being jammed); the distance to the jammer and to the target are the same, and the jamming enters the radar's antenna at the same angle as the desired signal (i.e., $D_J = D_T$ and $G_{T/R} = G_{RJ}$). The J/S formula reduces to:

$$J/S \text{ (in dB)} = 71 + P_J - P_T + G_J - G_{T/R} + 20 \log (D_T) - 10 \log(\sigma)$$

Consider the same radar and target, but put the jammer on the platform being tracked by the radar and reduce its power to 100W and its antenna gain to 10 dB. Now the J/S is calculated to be:

$$J/S = 71 + 50 \text{ dBm} - 60 \text{ dBm} + 10 \text{ dB} - 30 \text{ dB} +$$
$$20 \log(10) - 10 \log(10) = 51 \text{ dB}$$

9.3 Burn-Through

Burn-through is a very important concept in jamming, since it deals with the operational circumstances under which jamming remains effective. Burn-through occurs when the J/S ratio is reduced to the point where the receiver being jammed can adequately do its job.

9.3.1 Burn-Through Range

Burn-through range is defined in terms of radar jamming but can also be applied to communications jamming. In radar jamming, the burn-through range is the distance to the target at which the radar has adequate signal quality to track the target. Figure 9.11 shows the burn-through range for both self-protection and stand-off jamming. In both cases, it refers to the range from the radar to the target.

In communications jamming, the concept of burn-through range is not quite so graphic, but it is still sometimes useful. In this case, burn-through range means the effective range of the communications link in the presence of a specific jamming application (see Figure 9.12). It is the transmitter-to-receiver distance at which the receiver has adequate signal-to-noise ratio to demodulate and recover the required information from the desired signal.

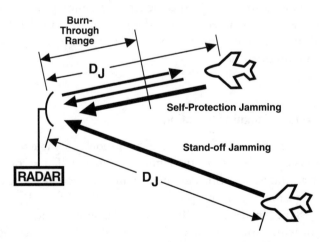

Figure 9.11   The burn-through range is the range from the radar to its target at which the jammer can no longer prevent the radar from doing its job.

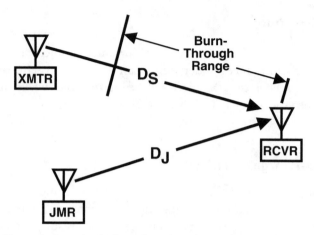

Figure 9.12 The equivalent of radar burn-through against communications jamming occurs when the range from the desired transmitter to the receiver is reduced to the point at which the signal is received with adequate quality.

9.3.2 Required J/S

The J/S ratio required for effective jamming can vary from 0 to 40 dB or more, depending on the type of jamming employed and the nature of the desired signal modulation. As the specific types of jamming are discussed, the required J/S will be included in each discussion. Because 10-dB J/S is a nice round number that applies in many situations, it will be defined as "adequate" in this discussion.

9.3.3 J/S Versus Jamming Situation

The J/S ratio varies with many parameters, as shown in Table 9.2. The first column shows every element of the jamming situation. The second column shows how an increase in this parameter will affect the J/S. For example, increasing the jammer transmit power increases J/S dB for dB, so doubling P_J doubles J/S (i.e., increases it 3 dB). The third column identifies the type of jamming for which each parameter is meaningful, and in one case (radar antenna gain) it differentiates the different effect on J/S. (The effect is significantly greater for stand-off jamming.)

Table 9.2
The Effect of Each Parameter in the Jamming Situation on J/S

Parameter (Increasing)	Effect on J/S	Type of Jamming
Jammer transmit power	Directly increases on J/S dB for dB	All
Jammer antenna gain	Directly increases J/S dB for dB	All
Signal frequency	None	All
Jammer-to-receiver distance	Decreases J/S as the distance^2	All
Signal transmit power	Directly decreases J/S dB for dB	All
Radar antenna gain	Decreases J/S dB for dB	Radar (self-protect)
Radar antenna gain	Decreases J/S 2 dB per dB	Radar (stand-off)
Radar-to-target distance	Increases J/S as the distance^4	Radar
Radar cross-section of target	Directly increases J/S dB for dB	Radar
Transmitter-to-receiver distance	Increases J/S as the distance^2	Comm
Transmit antenna gain	Directly decreases J/S dB for dB	Comm
(Directional) receiver antenna gain	Directly decreases J/S dB for dB	Comm

9.3.4 Burn-Through Range for Radar Jamming (Stand-Off)

The formula for burn-through range for each kind of jamming is just the appropriate J/S equation presented with all terms defined in the previous section, but rearranged to isolate the range term. (Remember that the constants in these convenient dB formulas dictate the units for the inputs and outputs—in this case, the range is in km.) The J/S formula for stand-off radar jamming is:

$$J/S = 71 + P_J - P_T + G_J - 2G_{T/R} + G_{RJ} - 20 \log(D_J) + 40 \log(D_S) - 10 \log(\sigma)$$

which can be rearranged:

$$40 \log(D_S) = -71 - P_J + P_T - G_J + 2G_{TR} - G_{RJ} + 20 \log(D_J) + 10 \log(\sigma) + J/S$$

The expression for 40 log (D_S) can be calculated from the various signal and jamming parameters. Since this is a radar, let's change D_S to D_T (the distance to the target). D_T is a dB number, which must be converted back to distance units (km in this case). The burn-through distance is:

$$D_T = 10^{\left(\frac{40 \log (D_T)}{40}\right)}$$

For example, consider the situation in which the jammer power is 1 kW (+60 dBm) into a 20-dB antenna; the radar has a 1-kW transmitter and a 30-dB antenna gain; and the jammer is 40 km away in a 0-dB antenna side lobe. The target radar cross-section is 10 m², and a 10-dB J/S is required for adequate jamming.

$$40 \log(D_T) = -71 - 60 \text{ dBm} + 60 \text{ dBm} - 20 \text{ dB} + 60 \text{ dB} - 0 \text{ dB} + 20 \log(40) + 10 \log(10) + 10 \text{ dB} = 21 \text{ dB}$$

$$D_T = 10^{(21/40)} = 3.3 \text{ km}$$

So the radar will not be able to track its target at any range greater than 3.3 km.

9.3.5 Burn-Through Range for Radar Jamming (Self-Protection)

The formula for self-protection J/S is:

$$\text{J/S (in dB)} = 71 + P_J - P_T + G_J - G_{T/R} + 20 \log(D_T) - 10 \log(\sigma)$$

It can be rearranged to read:

$$20 \log(D_T) = -71 - P_J + P_T - G_J + G_{T/R} + 10 \log(\sigma) + \text{J/S}$$
$$\text{and:}$$

$$D_T = 10^{\left(\frac{20 \log (D_T)}{20}\right)}$$

For example: Consider the situation in which the self-protection jammer power is 100W (+50 dBm) into a 10-dB antenna and the radar has a 1-kW transmitter and a 30-dB antenna gain. The target radar cross-section is 10 m², and a 10-dB J/S is required for adequate jamming.

$$20 \log(D_T) = -71 - 50 \text{ dBm} + 60 \text{ dBm} - 10 \text{ dB} +$$
$$30 \text{ dB} + 10 \log(10) + 10 \text{ dB} = -21 \text{ dB}$$

$$D_T = 10^{(-21/20)} = 89\text{m}$$

so the target aircraft can protect itself from tracking by this radar clear in to 89m.

9.3.6 Burn-Through Range for Communications Jamming

The dB formula for J/S in communications jamming is:

$$\text{J/S (in dB)} = P_J - P_T + G_J - G_T - 20 \log(D_J)$$
$$+ 20 \log(D_S) + G_{RJ} - G_R$$

which can be rearranged as:

$$20 \log(D_S) = -P_J + P_T - G_J + G_T + 20 \log(D_J) - G_{RJ} + G_R + \text{J/S}$$
$$\text{and:}$$
$$D_S = 10^{\left(\frac{20 \log (D_T)}{20}\right)}$$

For example: Consider the situation in which the jammer's transmitting power is 100W (+50 dBm), its antenna gain is 10 dB, and its distance from the receiver is 30 km—the desired signal transmit power is 1W (+30 dBm), its antenna gain is 3 dB, and the receiver's antenna provides 3-dB gain to both the desired and jamming signals. A 10-dB J/S is required.

$$20 \log(D_S) = -50 \text{ dBm} + 30 \text{ dBM} - 10 \text{ dB} + 3 \text{ dB} + 20 \log(30)$$
$$- 3 \text{ dB} + 3 \text{ dB} + 10 \text{ dB} = 13 \text{ dB}$$

$$D_T = 10^{(13/20)} = 4.5 \text{ km}$$

which means that the communications link being jammed will be able to function up to a distance of 4.5 km against the jamming.

9.4 Cover Jamming

The last two sections have classified jamming as either "stand-off" or "self-protection." Two other important classifications are "cover" and "deceptive."

Cover jamming, which typically uses noise modulation, simply decreases the signal-to-"noise" ratio in the jammed receiver as much as possible. Deceptive jamming causes a radar to draw false conclusions about the location or velocity of the target it is attempting to track. This section's focus is cover jamming, including the concept of power management to maximize jamming effectiveness.

Every type of receiver must have an adequate signal-to-noise ratio in order to properly process the signals it is designed to receive. The SNR is the power ratio of the desired signal to the noise power in the receiver's bandwidth. In a nonhostile environment, the noise power is the thermal noise of the receiving system [i.e., kTB (in dBm) + the receiver system noise figure (in dB)]. The received desired-signal power is a function of the transmitter power, the length of the transmission path, the operating frequency, and (for radars) the RCS of the target. Cover jamming injects additional noise into the receiver, which has the same effect as increasing the transmission-path length or decreasing the RCS of a radar's target.

When the jamming noise is significantly higher than the receiver's thermal noise, we speak of the J/S ratio rather than the SNR, but the effect on signal reception and processing is the same. If cover jamming is increased gradually, the operator or the automatic processing circuitry following the receiver may never become aware that jamming is present—only that the "SNR" is becoming extremely low.

The required RCS depends on the nature of the received signal and the way it is processed to extract its information. For voice communications, the SNR will depend on the skill of the speaker and the listener and the nature of the messages being passed. Effective communication ceases when the SNR rises to the point at which no information can be received. For digital signals, inadequate SNR causes bit errors and communication ceases when the bit error rate is too high to pass messages.

For radar signals, a skilled operator can usually manually track a single target at a much lower SNR than that required for an automatic tracking circuit that handles multiple targets. Thus, the goal of radar jamming may be to defeat the ability of the radar to track automatically—that is, making the radar reach saturation with many fewer targets.

9.4.1 J/S Versus Jammer Power

As shown in Figure 9.13, a receiving system discriminates to some extent against all signals but the one it is designed and controlled to receive. If it has a directional antenna pointed at the source of the desired signal, all signals from other directions are reduced. Any type of filtering (bandpass filters,

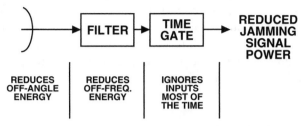

Figure 9.13 The only part of the transmitted jamming signal that is effective is that which penetrates all of the radar's angle, frequency, and timing selectivity, which are designed to optimize the reception of its return signal.

tuned preselection filters, IF filters) reduces out-of-band signals. In pulsed radars, the processor following the receiver knows approximately when to expect a return pulse and will ignore signals that are not near the expected return time.

If frequency hopping is employed in either radar or communications applications, the frequency band accepted by the receiver is a "moving target." When other types of spread-spectrum techniques are used, the signal is spread over a wide frequency range that the receiver can reverse to achieve the sensitivity appropriate to the signal before it was spread.

The problem for the jammer is that to be effective, it must spread its available power over the entire frequency that the receiver *might* be receiving—over all the angular space that *might* contain the receiving antenna—during all of the time that the receiver *might* be accepting signal

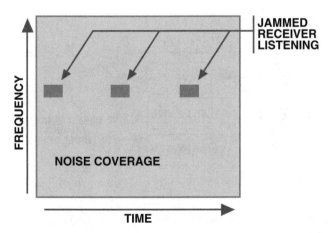

Figure 9.14 Noise-jamming energy must be spread over the total time-frequency space in which the receiver's desired signal might be present.

energy. Still, it is only the amount of power that gets through all of the receiver's defenses, as shown in Figure 9.14, that contributes to J/S. Since a jammer's transmitter power is directly related to its size, weight, prime power availability, and cost, the answer is seldom just to increase the jammer output power until there is enough *effective* jammer power.

9.4.2 Power Management

The more the jammer knows about the operation of the receiver, the more narrowly it can focus its jamming power to what the receiver will notice. Jammer energy-focusing is called "power management," and it can only be as good as the information available about the jammed receiver. This information normally comes from a supporting receiver (either a jammer receiver or an electronic support system), which receives, classifies, and measures the parameters of signals that are thought to be received by the receiver being jammed. Sometimes this is easy (as in a radar that is tracking the platform carrying the jammer) and sometimes this is harder (for example, communications links or bistatic radars). The integrated EW system shown in an extremely simplified block diagram in Figure 9.15 will provide its jammer with information on direction of arrival, frequency, and timing appropriate for managing its power.

The bottom line is that the jammer can concentrate its power where it will do the most good. As shown in Figure 9.16, power management will also reduce the jamming platform's vulnerability to home-on-jam threats by reducing the transmitted jammer power available to them.

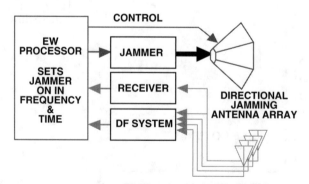

Figure 9.15   A power management system focuses the jamming power in the direction, frequency, and time slots occupied by the radar return signal with as little wasted energy as possible.

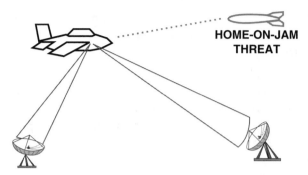

Figure 9.16 Directing jamming energy toward receivers to be jammed both increases jamming effectiveness and provides reduced vulnerability to home-on-jam weapons.

9.4.3 Look-Through

In order for power management to be effective, it is necessary to continue to receive the signals containing the information about the receiver being jammed. This process is called "look-through" and is most directly accomplished by stopping the jamming for brief periods to allow the look-through receiver to "take a peek." A spirited discussion about look-through periods persists among the receiver and jammer experts developing integrated EW systems. The jamming gaps must be long enough for the receiver to find and measure signals, yet short enough to leave adequate jamming effectiveness—a hard problem! (Please note that look-through is also covered—from the receiver's point of view—in Chapter 6.) There are several other techniques for isolating the receiver from the jamming signal that can be used with or instead of traditional look-through:

- *Antenna isolation.* The jammer has a narrow antenna beam with significantly reduced power in the direction of the supporting receiver. In some cases, the receiver has a narrow-beam antenna that adds isolation, but this doesn't work in systems requiring continuous omni-directional coverage. If the jamming antenna can be cross-polarized to the receiving antenna, significant additional isolation can be received.

- *Physical separation of the jammer from the supporting receiver.* This is achievable on a single large platform or via separate receiving and jamming platforms. The separation isolation can be increased through the use of radar-absorptive material or by careful spacing to take advantage of phase-related fading phenomena. If there is much distance between the jammer and receiver, coordination can become challenging.

- *Phase cancellation.* This occurs through the insertion of a phase-reversed version of the jamming signal into the receiver input. This is challenging because the receiver will observe the jamming as a complex combination of multipath signals with complex and changing phase characteristics.

9.5 Range Deceptive Jamming

The next few sections will discuss various deceptive jamming techniques. Deceptive jamming is almost entirely a concept applicable to radars. Rather than reducing the signal-to-noise ratio in the receiver, these techniques operate on the radar's processing to cause it to lose its ability to track a target. Some cause the radar track to move away from the target in range, and some in angle. First, we will discuss techniques which do not work against monopulse radars (i.e., radars in which each pulse contains all necessary tracking information). Then we will cover the monopulse jamming techniques. The first deceptive techniques discussed are "range gate pull-off" and the related "inbound range gate pull-off."

9.5.1 Range Gate Pull-Off Technique

This is a self-protection technique that requires knowledge of the time of arrival of pulses at the target being tracked by the radar. The jammer emits a false return pulse that is delayed from the reflected radar pulse by a gradually increasing amount, as shown in Figure 9.17. Since the radar determines the range to the target by the time of arrival of reflected pulses, this technique makes the radar "think" that the target is farther away than it actually is. The effect is to deny the radar accurate range information. This technique requires 0- to 6-dB J/S ratio.

As shown in Figure 9.18, a radar tracks the target in range by use of early and late gates. When the pulse energy in one gate becomes greater, the radar moves both gates to equalize the energy, thus tracking the target in range. By adding a stronger pulse over the true return pulse, the jammer "captures" the gates and creates enough jamming pulse energy to pull the gates away from the true or "skin" return arrival times.

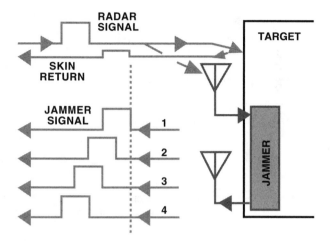

Figure 9.17 The range gate pull-off jammer transmits a higher-power return signal and delays it by an increasing amount.

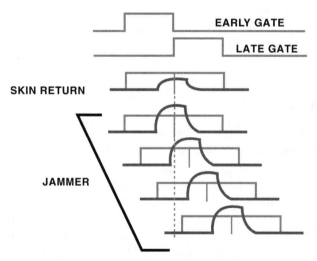

Figure 9.18 The jammer adjusts the timing of its early and late gates to balance the higher jammer pulse power.

9.5.2 Resolution Cell

A radar has a "resolution cell" in which it can resolve a target. The range dimension of the resolution cell is normally taken to be half the length of the pulse in range (i.e., the distance equal to half the pulse duration multiplied by the speed of light); the cell width is normally considered to be the radar

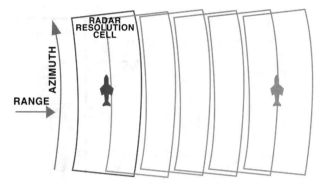

Figure 9.19   The range gate pull-off jammer pulls the radar's resolution cell away from the target in range, but the azimuth remains accurate.

antenna beamwidth (i.e., twice the sine of half of the antenna's 3-dB beamwidth multiplied by the range from the radar to the target). The process of tracking a target can be thought of as trying to keep the target centered in a resolution cell. By moving the range gate out in time, the range gate pull-off jammer moves the resolution cell away from the target, as shown in Figure 9.19. When the real target is outside the resolution cell, the radar track has been broken.

9.5.3   Pull-Off Rate

An important consideration is how fast the jammer can pull the range gate away from the target. Obviously, the faster the range gate is moved, the better the protection. However, if the pull-off rate exceeds the radar's tracking rate, the jamming will fail. If you don't know anything about the design of the radar being jammed, you can set this limit by considering the job the radar is designed to do. The radar must be able to track the maximum rate of change of the target range (i.e., the target moving directly toward or away from the radar), and to change its range tracking rate at the maximum rate of change of the range rate (i.e., the range acceleration).

9.5.4   Counter-Countermeasures

Two counter-countermeasures are effective against range gate pull-off jamming. One is simply to increase the radar's power so that the true skin return dominates the return signal tracking. This is in effect what happens at the "burn-through" range. The second is to use leading-edge tracking. Consider the actual signal received by the radar during range gate pull-off

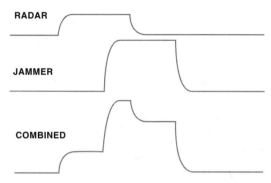

Figure 9.20   The combined jammer and return signals to the radar receiver include the pulse information from both signals.

jamming. As shown in Figure 9.20, both the skin-return and jammer pulses are present, and with adequate resolution, you can discern the leading and trailing edges of both pulses.

By differentiating the combined return signal, the radar would see a signal as shown in Figure 9.21, with spikes at the leading edges of the two pulses. If the radar is tracking this leading-edge signal, it will not be pulled away as the jammer pulse's leading edge moves later in relative time.

9.5.5   Range Gate Pull-In

A related jamming technique overcomes leading-edge tracking by pulling the range gate toward the radar rather than away from it. This technique is called "inbound range gate pull-off," or simply "range gate pull-in." Figure 9.22 shows the movement of the jamming pulses in this technique. The leading edge of the jamming pulse will now precede the leading edge of the skin

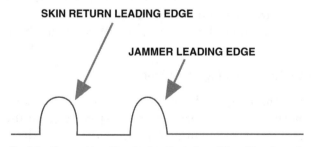

Figure 9.21   By detecting and tracking the leading edge of the skin return signal, the radar can remain locked to the skin return.

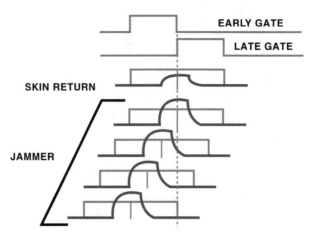

Figure 9.22 By anticipating the pulse from the radar, a range gate pull-in jammer can defeat leading-edge tracking, but this becomes very difficult if the radar does not have a single stable pulse-rate frequency.

return pulse, so it can steal the leading edge tracker. It is necessary to know the pulse repetition interval (PRI) to predict the arrival time of the next pulse in a pulse train, so that the jammer pulse can precede it by a carefully controlled amount. Thus, range gate pull-in is fairly easy against a radar with a single PRI. However, this technique requires great sophistication for employment against staggered pulse trains and cannot work at all against randomly timed pulses.

9.6 Inverse Gain Jamming

Inverse gain jamming is one of the techniques used to cause a radar to lose angular track. If successful, this technique will either deny angle-tracking data to the radar's processor or cause it to make improper tracking-correction commands when reacting to the combination of the skin-return signal and the jamming signal. This technique requires a 10- to 25-dB J/S ratio.

9.6.1 Inverse Gain Jamming Technique

Inverse gain jamming is a self-protection technique that uses the radar's antenna-scan gain pattern as seen by a receiver at the target being illuminated. Figure 9.23 shows a typical radar scan pattern. As the radar beam sweeps by the target, the time history of the power it applies to the target

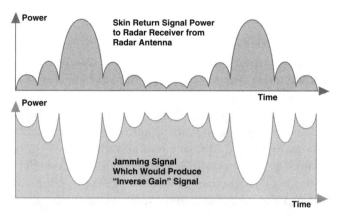

Figure 9.23 An ideal inverse gain jammer produces a signal that represents the inverse of the gain of the radar's receiving antenna so the radar receiver receives a constant signal level.

varies, as shown in the top part of the figure. This is called the *threat radar scan*. The large lobes occur as the radar's main beam passes the target, and the smaller lobes occur as each radar side lobe passes the target. The skin return from the illuminated target is reflected back to the radar with this same scan pattern, and the radar uses the same antenna to receive the return signal. Basically, a radar determines the angle (the azimuth, elevation, or both) to the target by knowing where its main beam is pointed when it receives the maximum skin-return signal strength.

If a transmitter located at the target were to transmit a signal back toward the radar with the same modulation (e.g., the same pulse parameters) as the radar, but with power versus time as shown in the bottom part of the figure, the received signal power and the radar's antenna gain would add to a constant. This means the receiver in the radar would receive a constant-strength signal, regardless of where its antenna beam is pointed — and would thus be unable to determine angular information about the target location.

Old EW hands will realize the above description is oversimplified in several ways, but this idealized case of inverse gain jamming illustrates the principle. Practical applications can vary from this ideal in several ways. One way is to apply jamming with the inverse gain pattern only during the period when the main beam is near the protected target. Several other implementations of this jamming technique use less artistic jamming waveforms.

9.6.2 Inverse Gain Jamming Against Con Scan

A con-scan radar scans its antenna beam in a circular motion (thereby describing a cone in space). Information from the scan is used to move the radar so that the target is in the center of the "cone." During tracking, the target is always within the antenna's main beam, but there is a sinusoidal variation in received power if the target is not centered in the cone. Figure 9.24 shows the shape of the antenna's main beam, the circular motion of the antenna, and the resulting threat-antenna scan pattern. As the antenna boresight (maximum gain direction) is moved around the circular path, the boresight is much closer to the target at point A than it is at point B. Thus, the radar antenna has greater gain in the direction of the target at A than at B, so more signal power is applied to the target at A than B.

Figure 9.25 illustrates the technique for implementing inverse gain jamming against a con-scan radar. The top line of this figure shows the sinusoidal amplitude pattern of the signal that reaches the target; this is also the shape of the skin-return signal received by the radar. By sensing the amplitude and phase of the skin return, the radar can move the center of its conical scan toward the target. The closer the target is to the center of the scan, the smaller the sinusoidal pattern. If the target is centered in the scan, the skin return will be at a constant power level—typically of the order of 1 dB less than that produced at the antenna boresight.

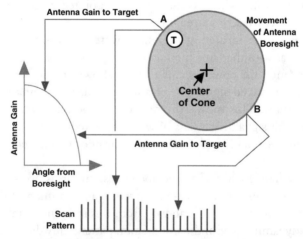

Figure 9.24   An antenna beam moved in a conical scan produces a sinusoidal output when observed at a target not centered in the scan.

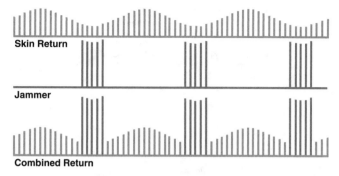

Figure 9.25 Bursts of strong, synchronized pulses from a jammer during the minimums of a con-scan waveform cause inverse gain jamming.

As shown on the second line of the figure, the jammer applies bursts of high-power pulses synchronized to the radar pulses. The period of the bursts is the same as the scan period of the radar antenna, and thus equals the period of the sinusoidal scan pattern. These bursts are timed to the minimums of the radar scan cycle received at the target. This means that the total signal received by the tracking radar will be as shown on the third line of this figure.

Now consider how the radar's tracking mechanism reacts to this combined signal. The antenna scan angle at which the *skin-return minimum* occurs is now the *maximum signal power* angle, so the tracker steers the radar scan directly *away from* the target rather than toward it. When successful, this causes the radar track to move far enough away from the target so that the radar track is "broken" and the radar must try to reacquire the target and begin the tracking process again.

9.6.3 Inverse Gain Jamming of TWS

Figure 9.26 shows the concept of a track-while-scan (TWS) radar using two fan beams. The two beams transmit (and receive) signals at different frequencies. One beam measures elevation of all targets observed and the other measures the azimuth—thus allowing the radar to simultaneously know the locations of multiple targets within the tracking range. This figure represents the angular space over which tracking takes place; range is measured by the time of arrival of reflected pulses.

As shown in Figure 9.27, the radar can determine the azimuth to the target by noting the position of the azimuth (or vertical) beam when the maximum skin return occurs. The position of the elevation (or horizontal) beam when it receives the maximum skin return determines the elevation of

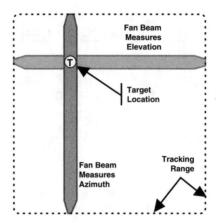

Figure 9.26   One significant type of TWS radar measures target azimuth and elevation using different beams.

the target. If the two beams happen to pass the target at the same time (as in Figure 9.26), the two responses would be time-synchronized.

Figure 9.28 illustrates the inverse gain jamming of the TWS radar. This figure considers only one of the beams, but the technique could be used against either beam or both. The first line is the skin return in a single beam. The radar will track the target in this beam by balancing the energy in the early and late gates of the angle gate. The second line of the figure is the jamming signal—bursts of pulses synchronized with the radar's pulses. The third line is the combined skin return and jamming signal received by the radar receiver. If the jammer pulse bursts are time-swept (either direction), they will move through the target return, capture the angle gate and thus cause the TWS radar to lose lock on the target.

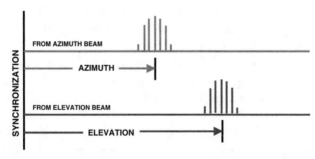

Figure 9.27   A TWS radar using the antenna beams as in Figure 9.26 measures target location as the time of skin returns in the two beams.

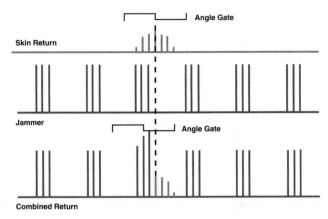

Figure 9.28   The inverse gain jammer pulls the angle gate in each beam away from the skin return.

9.6.4   Inverse Gain Jamming of SORO Radars

A scan-on-receive-only (SORO) radar illuminates the target with a steady signal from an antenna that follows the target. It uses tracking information from a receiving antenna that is scanning. As shown in Figure 9.29, a receiver located at the target would see a signal of constant amplitude, so the jammer could not measure the period of the radar scan or determine the location of the minimums. However, if the receiver at the target can identify the type of radar in use, it will know the approximate scanning rate. Figure 9.30 shows the way that inverse gain jamming is applied to the SORO radar. The first line of the figure shows the received skin return signal (since the shape of the tracking pattern is caused by the receiving antenna scan, this waveform

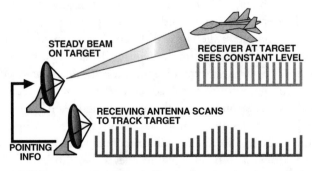

Figure 9.29   A SORO radar applies a steady illumination signal to the tracked target, tracking it with a scanning receive antenna.

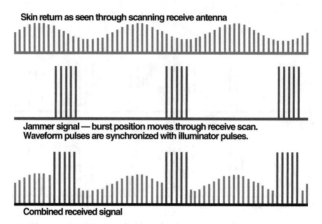

Figure 9.30   Moving periodic bursts of synchronized pulses through the tracking waveform of a SORO radar causes inverse gain jamming.

only exists inside the radar). As shown on line two of the figure, the jammer produces bursts of pulses synchronized with the radar pulses. The burst rate is slightly above or below the assumed scanning rate of the receiving antenna, causing the bursts to "walk through" the scan pattern of the receiving antenna, as shown on the third line of the figure. Although this jamming burst pattern will not consistently create 180° tracking errors (as it would if it were synchronized to the scan of the jammed radar), it will cause erroneous tracking signals almost all of the time.

9.7 AGC Jamming

Automatic gain control (AGC) is an essential part of any receiver that must handle signals over an extremely wide received-power range. The receiver's instantaneous dynamic range is the difference between the strongest and weakest signals it can receive simultaneously. To accept a range of signals wider than this *instantaneous* dynamic range, it must incorporate either manual or automatic gain control, to reduce the level of all received signals enough to allow the strongest signal to be accepted. AGC is implemented by measuring power at some appropriate point in the receiving system and automatically reducing the system gain or increasing an attenuation enough to reduce the strongest in-band signal to a level that can be handled by the receiver.

Large variations in target range and RCS require the use of AGC in radars. (If you are short of something fun to do, you are invited to go to

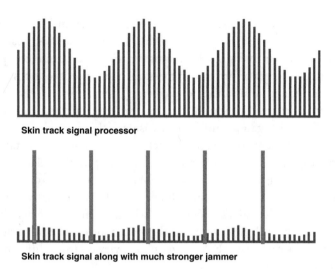

Skin track signal processor

Skin track signal along with much stronger jammer

Figure 9.31 The AGC jammer captures the radar's AGC, reducing its tracking signals to prevent angle tracking.

Section 2.5 and compare the power to the radar receiver from a 0.1-m² RCS at 100 km to that from a 200-m² RCS at 1 km.) Since a radar receiver is designed to receive only one signal (i.e., the skin return of its transmitted signal), it has little need for a wide instantaneous dynamic range, but must be able to quickly reduce its gain to accept a large skin return signal. It must then hold that reduced gain setting while it makes the relatively precise amplitude measurements required to track the target. Hence, the radar has a fast attack/slow decay AGC.

An AGC jammer transmits very strong pulses at approximately the radar's antenna scan rate. As shown in Figure 9.31, these pulses capture the radar's AGC. The resulting gain reduction causes all in-band signals to be greatly reduced. The skin track signal is suppressed to such a low level that the radar cannot effectively track the target.

9.8 Velocity Gate Pull-Off

Continuous wave (CW) and pulse Doppler (PD) radars separate signals reflected by a moving object (for example, a low-flying aircraft or a walking soldier) from signals reflected by the Earth using frequency discrimination. According to the Doppler principle (see Chapter 8), the reflected radar return signal from everything within the radar's antenna beam will be changed in frequency. The frequency shift of the reflection from each object is proportional to the relative velocity of the radar and the object causing the reflection. As shown in Figure 9.32, this return can be quite complex. To track a particular target return in this mess, the radar needs to focus on a narrow frequency range around the desired return signal. Since every frequency in the Doppler return corresponds to a relative velocity, this frequency filter is called a "velocity gate," and is set to isolate the desired target return. During an engagement, the relative velocity of the radar and the target may change rapidly and over a wide range—for example, the relative velocity between two aircraft making 6-g turns at Mach 1 can range from Mach 2 to 0 and change at rates of up to 400 kph per second. As the relative velocity of the target changes, the radar's velocity gate will move in frequency to keep the desired return centered. Also note that the amplitude of the return signal can change rapidly because the RCS of any object viewed from different angles can differ significantly.

The operation of a velocity gate pull-off (VGPO) jammer is described in Figure 9.33. The drawing shows in Figure 9.33(a) the target skin return centered in the velocity gate; none of the other elements present in a real-life return is shown. In Figure 9.33(b), the jammer generates a much stronger signal at the same frequency at which the radar signal is received at the target. The skin return will arrive back at the radar with a different frequency (Doppler shift), but since the target and the jammer move together, the jammer signal will be identically shifted and so will fall within the radar's velocity gate. In Figure 9.33(c), the jammer sweeps the jamming signal away from the frequency of the skin return. Since the jammer is much stronger, it

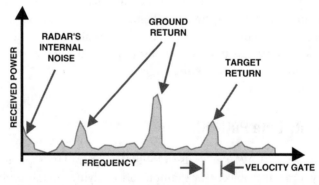

Figure 9.32 The AGC jammer captures the radar's AGC, reducing its tracking signals to prevent angle tracking.

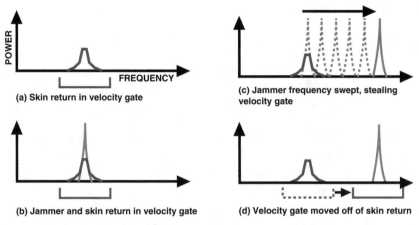

Figure 9.33   A velocity gate pull-off jammer uses the same principle as a range gate pull-off jammer, but in the frequency domain.

captures the velocity gate away from the skin return. In Figure 9.33(d), the jammer has caused the velocity gate to move far enough from the skin return that the skin return is outside the gate—breaking the radar's velocity track.

An important consideration is how fast the velocity gate can be pulled by the jammer. The answer depends on the design of the radar tracking circuitry, but a safe answer comes from the assumption that the radar is designed to be able to track a known class of targets. A study of the geometry of any type of EW engagement normally shows that the highest relative acceleration comes from turning rather than linear acceleration—so maximum target turning rates will give a good indication of the maximum rate of change of velocity that a radar must be able to follow.

9.9   Deceptive Techniques Against Monopulse Radars

The jamming of monopulse radars presents significant challenges. The deceptive techniques discussed so far won't work against them, particularly in self-protection jamming, and some jamming techniques actually enhance monopulse radar tracking. If a sufficient J/S ratio can be achieved by a stand-off jammer, it is effective against monopulse radar—as are properly deployed decoys and chaff if they produce adequate radar cross section. Decoys will be discussed in Chapter 10; the focus here is on deceptive techniques, which may be the best (or only) solution, depending on the tactical situation.

9.9.1 Monopulse Radar Jamming

A monopulse radar is tough to jam because it gets all of the information required to track a target (in azimuth and/or elevation) from each return pulse it receives, rather than by comparing the characteristics of a series of pulse returns. Self-protection jamming of monopulse radars gets even trickier, because the jammer is located on the target—a beacon that may make tracking even easier. If a self-protection jammer denies a monopulse jammer range information (with, for example, cover pulses), the radar can usually still track in angle, which may provide enough information to guide a weapon to the target.

There are two basic approaches to deceiving monopulse radars. One is to take advantage of some known shortcoming in the way the radar operates. The second is to take advantage of the way monopulse radars develop their angle-tracking information within a single radar resolution cell. The second approach is generally superior, so we'll discuss it first.

9.9.2 The Radar Resolution Cell

In Section 9.5 we briefly discussed the resolution cell—an area described by the radar beamwidth and pulse width. It deserves a more detailed treatment at this point—first the "width" of the cell, then the "depth" of the cell, as shown in Figure 9.34.

The width of the resolution cell is defined by the area that falls within the antenna's beam—which depends on the beamwidth and the distance

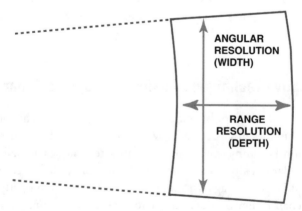

Figure 9.34 The width of the radar resolution cell is determined by the radar antenna beamwidth and the depth by the pulse duration.

from the radar to the target. The beamwidth is usually considered to be the 3-dB beamwidth, so at a range of n km the beam "covers" ($2n \times$ the sine of half the 3-dB beamwidth) km—but that doesn't tell the whole story. The radar's ability to discriminate between two targets in azimuth or elevation depends on the relative strength of the radar returns from both targets as the antenna beam is scanned across them. Clearly, if the targets are far enough apart that both cannot be in the antenna beam at the same time, the radar can discriminate between them (i.e., resolve them). Since the radar can usually be assumed to have the same transmitting and receiving antenna patterns, the return from a target located at the 3-dB angle from the antenna boresight will be received with 6 dB less power than a target at the boresight, as shown in Figure 9.35 (3 dB less power is transmitted, and the return is reduced by 3 dB).

Now consider what happens to the total received signal power from two targets separated in power by one-half beamwidth as the radar's antenna moves from one to the other. The power from the first target will diminish more slowly as the power from the second target builds up—so the radar will see one continuous "bump" of return power. At less than one-half beamwidth separation, this is even more pronounced. When the two targets are separated by more than a half beamwidth, the response has two "bumps," but they do not become pronounced until the targets are about a full beamwidth apart. Thus, the resolution cell can be considered to be a full beamwidth wide, but considering it to be one-half beamwidth across is more conservative.

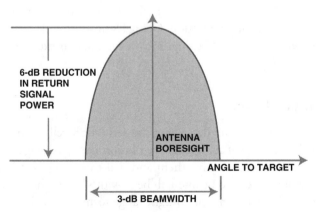

Figure 9.35   A radar target one-half beamwidth from the antenna boresight will produce 6-dB reduced return power.

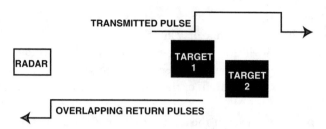

Figure 9.36 Two targets separated in range by one pulse duration will produce returns separated by a pulse duration.

The mechanism causing the depth of the resolution cell (i.e., the range resolution limitation) is shown in Figure 9.36. This figure shows a radar and two targets (with the distance to the targets obviously too short relative to the pulsewidth, PW). When the two targets are separated in range by less than half a pulse duration, the illumination of the second target starts before the illumination of the first target is complete. However, the arrival of the return pulse from the second target is delayed from the first return by twice the target separation divided by the speed of light—because the round-trip time from the range of the first target to the range of the second target is added. Thus, as the separation of the two targets in range is decreased, the return pulses do not start to overlap until the range difference reduces to half of a pulse duration—limiting the depth of the resolution cell to half a PW (in distance).

From the above discussion, the definition of the radar resolution cell is the area enclosed by the beamwidth and half the distance traveled by the radar signal during its pulse duration. George Stimson's book, *Introduction to Airborne Radar* (SciTech Publishing, 1998), includes an excellent, in-depth discussion of these points.

9.9.3 Formation Jamming

Having spent so much time describing the resolution cell, it suffices to say that if two aircraft are within a single resolution cell, as in Figure 9.37, the monopulse radar cannot resolve them and will thus track to the centroid. Taking the resolution cell to be one-half beamwidth by one-half PW, the two target aircraft will have to maintain tight range formation if the PW is short (e.g., 15m for a 100-nsec PW). The cross-range formation accuracy is more forgiving (e.g., 261m at 30-km range for a 1° radar beamwidth). Of course, the cell narrows significantly as the range diminishes.

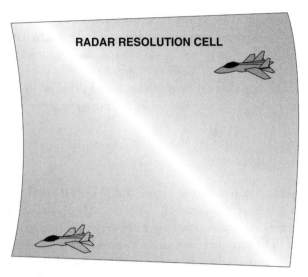

Figure 9.37 Formation jamming is achieved when two targets remain within a single radar resolution cell.

As shown in Figure 9.38, formation jamming can be performed at greater target range separations if cover pulse or noise jamming is used to deny the radar range information. The required J/S for these types of jamming is typically not high (0 to +10 dB).

9.9.4 Blinking Jamming

Blinking jamming also involves two targets within a single radar resolution cell. However, they carry jammers which are used cooperatively. The two

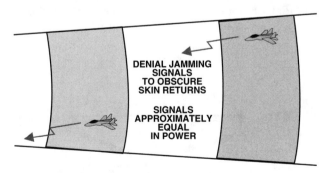

Figure 9.38 Formation jamming can be achieved with wider range separation of targets if the radar is denied range information.

jammers are activated alternately with a coordinated "blinking" rate that is close to the radar's guidance servo bandwidth (typically 0.1 to 10 Hz). If a resonance in the tracking response can be found, large antenna pointing overshoots can result. A missile guided toward a pair of properly blinking jammers will be guided alternately to one and the other with the swings becoming wilder as the range target diminishes, preventing proper terminal guidance.

9.9.5   Terrain Bounce

The terrain-bounce technique (see Figure 9.39) is particularly powerful against active or semi-active missile-guidance systems. A strong simulated radar return is generated and directed at an angle that will cause a reflection from the ground. The jammer transmission must have sufficient effective radiated power (ERP) to reflect a signal from the ground that reaches the missile-tracking antenna with significantly more signal strength than the skin return from the aircraft being attacked. If done properly, this will cause the missile to be guided below the protected aircraft.

9.9.6   Skirt Jamming

Figure 9.40 shows the amplitude pass band of a bandpass filter. Filters are designed to pass all frequencies within a pass band with as little attenuation as possible, while causing as much attenuation as possible to all signals outside the pass band. An ideal filter (sometimes called a "stone wall filter") would provide infinite attenuation to any signal even a tiny bit out of band. However, real-world filters have "skirts" in which input signals are attenuated by an amount proportional to the amount by which they are out of band. The slope of the skirt is 6 dB per octave—that is, the attenuation increases by a factor of four for each doubling of the frequency "distance" from the center of the filter's pass band—for each stage of filtering. Filters also have an "ultimate rejection" level, a maximum attenuation applied to signals that are far

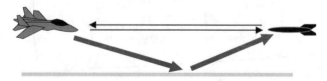

Figure 9.39   A strong retransmission of the radar signal aimed to bounce off of the ground will cause the radar tracker to guide to a point below the protected aircraft.

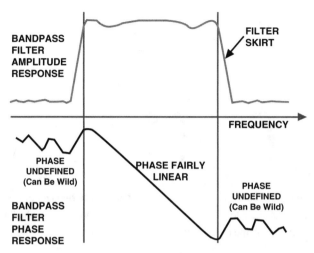

Figure 9.40 The filter amplitude response attenuates signals beyond the filter's bandpass, but the attenuation increases to a maximum across the filter's "skirt." The filter's phase response is undefined outside the passband.

out of band. The ultimate rejection is often about 60 dB. This means that a very strong out-of-band signal can get through the filter with some rejection if it is very close to the pass band and more if it is far out of band.

The other curve in Figure 9.40 shows the phase response of the filter. Across the pass band, a well designed filter will usually have a fairly linear phase response. However, beyond the band edges, the phase response is undefined and can be extremely nonlinear. This means that if a strong jamming signal is received in the "skirt" frequency range, it will have an erroneous phase, causing the radar's tracking circuitry to malfunction. The J/S ratio must, of course, be very high, because the jammer must overcome the filter's rejection and still have significantly more power than the true skin return.

9.9.7 Image Jamming

Figure 9.41 is a frequency spectrum diagram. As you will recall from Chapter 4, a superheterodyne receiver uses a local oscillator (LO) to convert an input RF frequency to an intermediate frequency (IF). This frequency conversion occurs in a mixer, which generates harmonics and the sums and differences of all signals input to it. The output of the mixer is filtered and passed to an IF amplifier (and then perhaps to another frequency-conversion stage). The frequency of the LO is either above or below the frequency of the

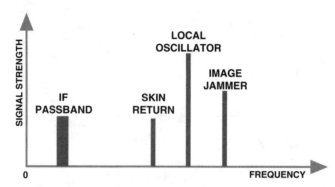

Figure 9.41 The intermediate frequency in a superheterodyne receiver or frequency converter is equal to the difference between the frequency to which the receiver is tuned and the frequency of the local oscillator.

desired receiver tuning frequency by an amount equal to the IF frequency. For example, in an AM broadcast receiver tuned to 800 kHz, the LO frequency is 1,255 kHz (because the IF frequency is 455 kHz). In this case, the "image" frequency is 1,710 kHz, and a signal received into the mixer at this frequency would also appear in the IF amplifier, causing severe degradation to receiver performance. To prevent such "image response," the receiver design almost always includes a filter to keep the image frequency away from the mixer.

Incidentally, the reason that wide-frequency-range reconnaissance receivers typically have multiple conversion designs is often to avoid image-response problems.

Assume for a moment that a particular radar receiver uses an LO whose frequency is above the frequency to which the receiver is tuned, as in Figure 9.41. The receiver is, of course, tuned to the appropriate frequency to receive the skin return; the IF frequency is equal to the difference between the skin-return frequency and the LO frequency. If a signal that looks like the skin return were received at the image frequency with enough power to overcome input filtering, it would also be amplified by the radar's IF amplifiers and processed along with the skin return. However, it would be reversed in phase from the true skin return, which would cause the radar's tracking error signal to change its sign (i.e., move the radar away from the target rather than toward it).

Unfortunately, this technique requires a great deal more knowledge about the radar's design than just its transmitted frequency (which will, of course, be the skin-return frequency without Doppler shifts). Does it use high-side or low-side conversion—that is, is the LO above or below the frequency of the skin return? If the radar receiver has little or no tuned front-end filtering, this technique requires only moderate J/S ratio, but if there is significant tuned filtering, 60 dB or more J/S may be required.

9.9.8 Cross-Polarization Jamming

Cross-polarization jamming can be effective against some radars that use parabolic dish antennas. The effectiveness is a function of the ratio of the antenna's focal length to its diameter, because the smaller this ratio, the greater curvature the antenna will have. If illuminated by a strong cross-polarized signal, the antenna will provide false tracking information because of cross-polarization lobes called "Condon" lobes. If the cross-polarization response is dominant over the matched polarization response, the radar tracking signal will change signs, causing the radar to lose track on the target.

To produce a cross-polarized signal, the jammer has two repeater channels with orthogonally oriented antennas (i.e., each with linear polarization, but 90° to each other) as shown in Figure 9.42. Although any set of orthogonal polarizations would work, they are shown as vertical and horizontal in the figure. If the vertically polarized component of the received signal is retransmitted with horizontal polarization, and the horizontal with vertical polarization, the received signal will be retransmitted cross-polarized to the received signal, as shown in Figure 9.43.

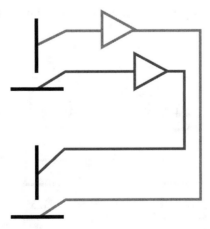

Figure 9.42 A cross-polarization jammer receives a radar signal through two orthogonally polarized antennas and retransmits each received signal with the orthogonal polarization.

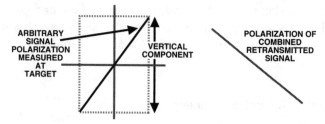

Figure 9.43   By retransmitting two orthogonally polarized signal components, each with 90° polarization shift, the cross-polarization jammer creates a signal which is cross-polarized to any linearly polarized signal received.

This technique requires 20–40 dB of J/S, depending on the design of the radar's antenna. It must be noted that an antenna protected by a polarization screen will have little vulnerability to cross-polarization jamming.

9.9.9   Amplitude Tracking

This is a good time to review the way that the tracking circuits in a monopulse radar provide target tracking. Consider a two-channel monopulse system, as shown in Figure 9.44. Two separate sensors (e.g., antennas) receive the skin-return signal. The angular tracking function is generated by a comparison of these two received signals. This requires that the difference between the two signals be emphasized to generate an error signal; the two received signals would be identical if a line connecting the two sensors were perpendicular to the target, but as the sensor-array boresight moves away, the

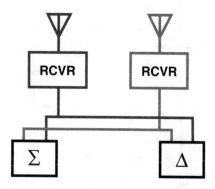

Figure 9.44   A monopulse tracker typically produces sum and difference signals for two receivers, then produces a tracking error signal from $\Delta - \Sigma$.

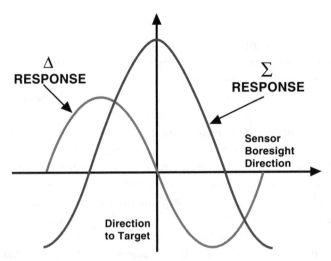

Figure 9.45   Sum and difference responses as a function of sensor boresight pointing relative to the tracked target.

tracker must generate an error signal that allows the array to move in the direction of the target. For proper target tracking, the error signal derived from the sensor outputs must be made independent of received signal strength, and the easiest way to do this is to normalize the difference signal by comparing it to a sum signal. Figure 9.45 shows the sum and the difference responses as a function of the angle between the boresight of the tracker and the direction to the target. (You will remember from our direction-finding discussions that the boresight of a set of sensors is perpendicular to a line connecting the two sensors.) To simplify the discussion, the sum signal will be identified with the symbol Σ and the difference signal with the symbol Δ. The tracking signal is generated from the quantity $\Delta - \Sigma$. The greater this value, the greater the magnitude of the correction the tracker must make to move its boresight to the target. The sign of Δ, of course, determines the direction of the correction.

9.9.10   Coherent Jamming

When two or more jammers are used together, they are coherent if the RF phase of the two jamming signals has a constant and controlled relationship. When the two coherent signals are in phase with each other, they will add constructively—but when they are 180° out of phase, they will cancel each other.

9.9.11 Cross Eye Jamming

Cross eye jamming involves a pair of repeater loops that are coherently related. Each retransmits a received signal from the position at which the other has received it. The separation of the two locations should be as great as possible. Figure 9.46 shows how a cross eye jammer might be implemented on the wingtips of an aircraft. Note that the two electrical paths must be identical in length, and one must have a 180° phase shift. To understand how the system works, we can revisit the concept of the "wavefront." As explained during the discussion of interferometric direction-finding (Chapter 8), the wavefront does not really exist in nature, but it is an extremely convenient concept. The wavefront is a line perpendicular to the direction from the transmitter. Since radio signals radiate spherically from an omnidirectional antenna (and behave more or less the same within the beamwidth of a directional antenna), the wavefront defines a line along which the phase of the radiated signal is constant.

Figure 9.47 shows that the total path length from the radar through the repeater and back to the radar remains identical for the two repeaters as the direction to the radar changes (as long as the repeater loops have identical lengths). The signals from the two repeaters will be 180° out of phase when they reach the radar's tracking antennas, independent of the direction to the radar. This causes a null in the combined response of the radar's sensors just

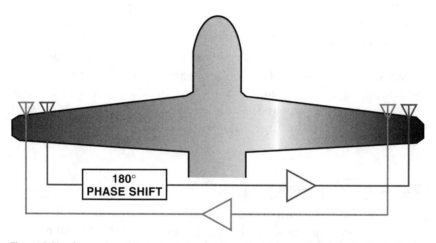

Figure 9.46 A cross eye jammer can be configured with two retransmission loops having antennas on the wing tips of an aircraft. One loop has a 180° phase shift. They are identical in electrical path length.

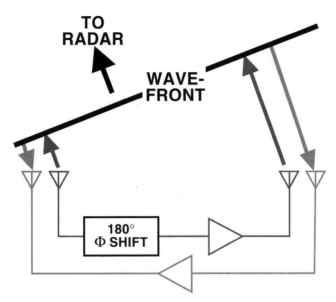

Figure 9.47 The electrical path length from the radar, through the jammer loop, and back to the radar is identical for the two paths independent of the direction from which the signal is received.

where the radar tracking circuit would expect a peak. Looking back at the sum and difference responses in Figure 9.45, we can see that if there is a null where there should be a peak in the sum response, the tracking signal will be greatly distorted.

This effect is often represented as a warping of the wavefront of the skin-return signal, as shown in Figure 9.48. This wavefront distortion is

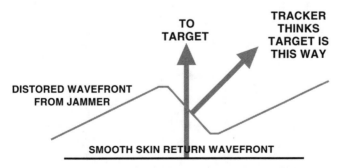

Figure 9.48 The cross eye jammer produces a discontinuity in the "wavefront" of the returned signal at the radar, causing it to generate a false tracking error signal.

repeated every few degrees. Note the center of the sharp discontinuity occurs right at the radar because of the effect shown in Figure 9.47.

Two important limitations impinge upon the application of the cross eye technique. One is the requirement that the electrical lengths of the two repeater paths be matched very closely (five electrical degrees is a common number). This is extremely difficult since the electrical "distance" through any circuitry, cabling or waveguide will vary with temperature and signal strength. (Remember, 5° is less than a millimeter at typical radar frequencies.) The technique's second restriction is its need for a very high J/S ratio (20 dB or more), because the null must overwhelm the sum signal.

10

Decoys

As the sophistication of guided weapons increases and particularly as "home-on-jam" modes gain wider use, the importance of radar decoys increases. In this chapter we will discuss the various types of decoys, their applications to the defense of military assets, and the strategies with which they deploy.

10.1 Types of Decoys

Decoys can be classified according to the way they are placed into service, the way they interact with threats, or the types of platforms they protect; terms abound for each category. To set some common vocabulary for the next few columns, we'll define the decoy *type* in terms of the way it is

Table 10.1
Missions and Platforms Typically Associated With Types of Decoys

Decoy Type	Mission	Platform Protected
Expendable	Seduction	Aircraft Ships
	Saturation	Aircraft Ships
Towed	Seduction	Aircraft
Independent maneuver	Detection	Aircraft Ships

223

deployed, the decoy *mission* in terms of the way it protects a target, and the *platform* as the military vehicle protected. Table 10.1 may spur dissent, since almost any type of decoy can be used in almost any type of mission to protect any kind of platform. If it hasn't been done yet, it probably will be in the very near future. At the moment, radar decoys described in the literature are limited to the protection of aircraft and ships. As millimeter-wave-radar guided munitions emerge to threaten surface-mobile targets, decoys will probably assume a ground-vehicle role as well.

Table 10.1, then, represents the major emphasis as described in current EW professional literature.

We'll divide the types of decoys into expendable, towed, and "independent maneuver." Expendable decoys are ejected from pods or launched in missiles from aircraft and launched from tubes or rocket launchers from ships. These decoys typically operate for short periods of time (seconds in the air, minutes in the water).

A towed decoy is attached to the aircraft by a cable, with which it can be controlled and/or retracted by the aircraft. Towed decoys are associated with long-duration operation. Towed barges, for ships, use large corner reflectors and could also be considered towed decoys, but they are typically considered separately.

Independent maneuver decoys are deployed on propelled, typically airborne, platforms. Examples are UAV decoy payloads, ducted fan decoys used in ship protection and decoys mounted on or below helicopters. When independent maneuver decoys protect a platform, they have complete flexibility of relative motion (in contrast to towed decoys, which must follow along; or expendable decoys, which either fall away or are fired forward). Ship protection is a prime application of independent maneuver decoys, as is forward penetration ahead of aircraft to expose enemy defenses for avoidance or attack.

10.1.1 Decoy Missions

Decoys have three basic missions: to saturate enemy defenses, to cause an enemy to switch an attack from the intended target to the decoy, and to cause an enemy to expose his offensive assets by preparing to attack a decoy. These three decoy missions are as old as the history of human conflict, far preceding the birth of electronic warfare. The difference is that rather than directly deceiving the senses of human warriors, modern EW decoys deceive the electronic sensors which detect and locate targets and guide weapons to them.

10.1.2 Saturation Decoys

Any type of weapon is limited in the number of targets it can engage at one time. Since a finite amount of time is allotted for the weapon's sensors and processors to deal with each of the targets it attacks, the limitation is more accurately described as a limit on the number of targets it can attack in a given amount of time. The total time period during which a weapon can engage a target starts when the target is first detected. It ends either when the target can no longer be detected or when the weapon has succeeded in performing its mission. The weapon will only be able to engage some maximum number of targets at once; if more targets are present, some will escape attack, because the weapon must operate above its saturation point.

A large number of decoys can be used to saturate a weapon or a combination of weapons—for example, an air-defense network. However, another variable comes into play with decoys. In general, the radar processing associated with a weapon system can either ignore or quickly discard the tracks of radar returns that are significantly different from the returns of intended targets. Thus, to be effective, decoys must look enough like real targets *to the weapon system's sensors* that they cannot be easily rejected. The more that is known about the sensors to be fooled, the more effective (and cost effective) the decoys can be. Ideally, the attributes of the decoy are *only* those that can be detected by the weapon system sensors; anything else adds size, weight, and cost. By the time an air-defense network processes all of the targets in Figure 10.1, the actual target may have accomplished its mission or may no longer be vulnerable to attack.

A special case of the saturation decoy mission occurs when the weapon system acquires a decoy first and then stops looking for the target (Figure 10.2). This is particularly important in protection against missiles with active guidance—for example, antiship missiles—which typically scan a narrow antenna beam to acquire the target ship after the missile breaks the horizon.

10.1.3 Detection Decoys

A new and particularly valuable use of radar decoys is to cause a defensive system, such as an air-defense network, to turn on its radars—making it susceptible to detection and attack. This typically requires independent maneuver decoys. If decoys look and act enough like real targets, the acquisition radars or other acquisition sensors will hand them off to tracking radars. Once the tracking radars turn on, they are vulnerable to attack by antiradiation

Figure 10.1   Saturation decoys force weapons sensors to deal with large numbers of apparent targets, reducing their ability to attack the real target.

missiles fired from aircraft outside the lethal range of the enemy weapon (Figure 10.3).

10.1.4   Seduction Decoys

In the seduction mission, the decoy attracts the attention of a radar that has established track on a target, causing the radar to change its track to the decoy. Then, the decoy moves away from the target, as shown in Figure 10.4. Tracking radars consider only narrow segments of azimuth (and sometimes elevation), range, and return-signal frequency—by use of angle, range, and frequency gates. If the decoy can move any or all of those gates far enough

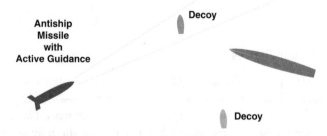

Figure 10.2   If a weapons sensor acquires a decoy before it detects a true target, it may attack the decoy, wasting an expensive guided missile.

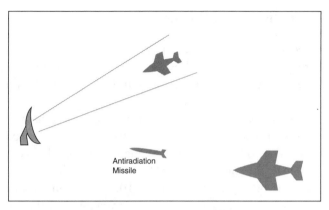

Figure 10.3 If a decoy forces an air-defense radar to track it, an attacking aircraft beyond the lethal range of the weapons system can attack it with an antiradiation missile.

away from the true target, the radar's tracking lock on the target will be broken. Thus, what we are calling seduction decoys could also be called "break-lock decoys."

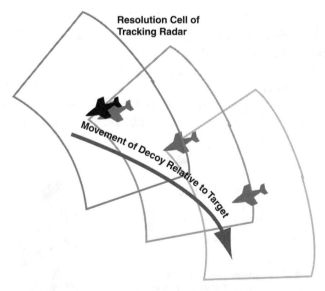

Figure 10.4 In the seduction mission, the decoy activates within the radar's resolution cell with the target, but with high apparent RCS. It captures the radar's tracking gates and moves them away from the target.

10.2 RCS and Reflected Power

Radar cross-section (RCS) is the effective reflecting area of anything which reflects radar signals. It is affected by the size, shape, material, and surface texture of the object causing reflection—and varies with frequency and aspect angle.

The important aspect of RCS in electronic warfare is the way it affects the reflected signal, because that gets directly into the signal part of the jamming-to-signal ratio (J/S). As shown in Figure 10.5, the RCS converts the illuminating power to reflected power. To someone with a "decoy point of view," the RCS can be as pictured in Figure 10.6. The "gain" associated with the RCS is the sum of the gains of the two antennas and the amplifier—remembering that any of those gains can be either positive or negative (i.e., losses). In the radar link discussion included in Chapter 2 an expression for the "gain" caused by reflection from the target is stated as:

$$P_2 - P_1 = -39 + 10 \log(\sigma) + 20 \log(F)$$

where P_2 is the signal leaving the target (in dBm); P_1 is the signal arriving at the target (in dBm); σ is the RCS (in m²); and F is the signal frequency (in MHz).

This expression, like all "dB equations," has to be considered with some qualifiers. First, understand that P_2 and P_1 are the reflected and illuminating power as they would be received by an ideal receiver with an omnidirectional antenna very close to the target reflecting the radar signal (but ignoring antenna near-field effects). As always, the constant (in this case −39) is put in to take care of the physical constants and the unit conversion factors. It is valid only if the proper units are used (in this case, dBm, m², and MHz).

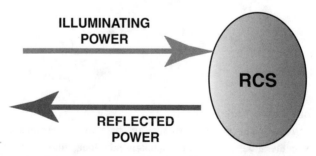

Figure 10.5   The RCS determines the ratio between the power illuminating a target and the power it reflects.

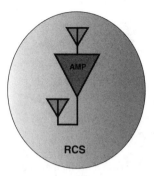

Figure 10.6 The RCS or either a target or a decoy can be considered as though it were an amplifier and two antennas. The effective signal gain caused by the RCS is the sum of the amplifier gain and the two antenna gains.

For example, a 10-GHz signal reflected by a 1 m² RCS will have a reflection "gain" of:

$$P_2 - P_1 = -39 + 10 \log(1) + 20 \log (10,000) = 41 \text{ dB}$$

Don't panic, the signal didn't increase its power when it was reflected from the target. This is the signal as received by that ideal receiver through its ideal omnidirectional antenna. If the RCS is high, it means that the reflected signal energy was effectively focused back toward the radar.

As an aside to help you understand why antennas are such a problem on "stealth" platforms: If you replace $P_2 - P_1$ with antenna gain (in dB) and σ with antenna effective area (in m²), you will have the equation relating the size of an antenna to its gain.

10.3 Passive Decoys

Passive decoys are merely radar reflectors. Their RCS is a function of their size and geometry, since they are always made of material which reflects radio energy well (typically metal, metalized fabric, or metalized fiberglass). Each simple shape has a characteristic maximum RCS. Since it is an extremely effective reflector and presents a high RCS over a fairly wide angle, the corner reflector is often used in passive decoys. As shown in Figure 10.7, an incoming signal is reflected three times to direct it back toward its source. Now consider the relative RCS of a cylindrical reflector and a corner reflector which would fit inside of it as shown in Figure 10.8. For realism, we consider the corner reflector with quarter-circular sides.

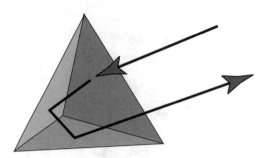

Figure 10.7 A corner reflector is extremely efficient, and provides retro-directive reflection over a wide angular range.

The maximum RCS of the cylinder is given by the formula:

$$\sigma = (2\pi\ a\ b^2)/\lambda$$

where a is the radius of the cylinder, b is its length, and λ is the signal wavelength. For σ in m², all of the lengths must be in meters.

The maximum RCS of the corner reflector (quarter circular sides) is:

$$\sigma = (15.59\ L^4)/\lambda^2$$

where L is the radius of the quarter-circular sides of the corner reflector.

The ratio of the two RCSs is:

$$\sigma CR/\sigma CYL = (15.59\ L^4\ \lambda)/(\lambda^2\ 2\pi\ a\ b^2)$$

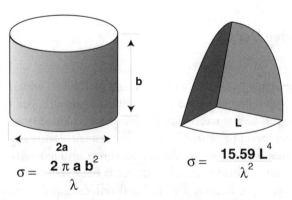

$$\sigma = \frac{2\pi a b^2}{\lambda}$$

$$\sigma = \frac{15.59\ L^4}{\lambda^2}$$

Figure 10.8 Placing a corner reflector inside a cylindrical area increases the RCS by a factor of more than 100 relative to a reflecting cylinder of the same size.

Letting both b and $L = 1.5$ a, $\sigma CR/\sigma CYL = 3.72$ L/λ.

As an example, if $L = 1$m and the frequency of the radar is 10 GHz, the corner reflector provides almost 75 times more effective cross-section or about 19 dB more reflected signal power.

10.4 Active Decoys

Referring back to Figure 10.6, the RCS can be considered to act like two antennas with an amplifier between them. The end-to-end gain of the two antennas and the amplifier equals the $P_2 - P_1$ signal gain caused by the RCS of the target.

If two real antennas and a real amplifier are used, they will have the same effect on the signal as an RCS which provided that same end-to-end gain. That is, in effect, how a physically small active decoy can simulate an RCS much larger than its physical size.

In practice, the decoy can actually use a "primed" oscillator to output a signal at some large, fixed power at the frequency of the received signal. In this case, the effective gain and the resulting equivalent RCS can be extremely large when the radar is distant from the target. However, as the radar approaches the target, the effective gain (hence "RCS") decreases.

Another active decoy implementation uses a "straight through repeater," which provides some fixed amount of gain for all received signals. Thus, the equivalent RCS remains the same as the radar approaches the target until the decoy's amplifier saturates—after which the equivalent RCS will decrease just as with the primed oscillator design.

An important consideration, particularly in physically small decoys, is that the two antennas must be sufficiently isolated from each other so that the transmitted return signal does not exceed the received signal in the receiving antenna at maximum end-to-end gain.

10.5 Saturation Decoys

Saturation decoys can be either passive or active, but they must provide RCS approximately equal to that of the target. They must also provide other characteristics detectable by the radar which are sufficiently close to those of the target to "fool" the radar. Examples of these characteristics are motion, jet engine modulation (if detected), and signal modulation. An example of a passive distraction decoy is shown in Figure 10.9. Here, chaff bursts (which provide passive RCS close to that of the protected ship) are output in a

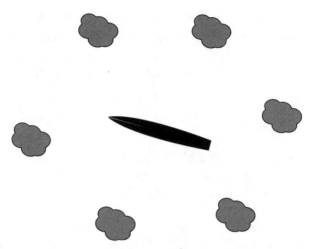

Figure 10.9 A ship in a pattern of chaff clouds with approximately the same RCS will cause an attacking missile to evaluate many targets to find the true target. This task is made more difficult by the fact that the ship is maneuvering and the chaff clouds are moving with the wind.

pattern so that an attacking radar controlled system must process every chaff cloud as though it were the target. (Please note that this drawing is not to scale; the chaff bursts would be put out in a much larger pattern.)

10.6 Seduction Decoys

Seduction decoys are so called because they "seduce" the tracking mechanisms of threat radars away from their intended targets. (And you thought you'd never find any sexual innuendoes in a technical book.) Decoys function in the seduction role after the threat radar has acquired the target. The purpose is to capture the tracking mechanism of the threat radar and break its lock on the target. Their function is much like that of a deceptive jammer (for example, a range gate pull-off jammer). However, the decoy is more powerful in that it holds the attention of the threat radar, which continues to track it. The range gate pull-off jammer, on the other hand, pulls the radar's range gate to a location which does not contain a target—allowing the radar to try to reacquire the target.

Another advantage of the decoy is, of course, that its signals are transmitted from a location away from the target. This defeats monopulse radars and home-on-jam modes.

10.6.1 Seduction Decoy Operating Sequence

As shown in Figure 10.10, the seduction decoy must turn on within the threat radar's resolution cell after the radar is tracking the protected target. To be effective, the decoy must return radar signals with sufficient power to simulate an RCS significantly larger than that of the protected target. For active decoys (Section 10.4), this requires adequate throughput gain and maximum power. For passive decoys (e.g., chaff bursts for ship defense), the effective decoy RCS must be greater than that of the target. Note that the RCS of the target is very much a function of the azimuth and elevation from which it is viewed, and maneuvering to reduce the target RCS presented to the attacking radar may be an integral part of the defensive strategy. It is also worth noting that the reduced RCS of modern "stealthy" platforms allows a better level of protection at any given decoy RCS.

As shown in Figure 10.11, the decoy captures the tracking mechanism of the threat radar, so the resolution cell moves to center itself on the decoy as it separates from the target. In this figure, the decoy is falling behind the target, but if the decoy is propelled, it could as well move away from the target in any direction. If the decoy seduction is successful (Figure 10.12), the decoy will pull the radar's resolution cell far enough away so that the protected target is completely out of the cell. At this point the effective J/S ratio of the decoy is infinite.

It is important to note that for the decoy to be effective, it must be indistinguishable from the target—as perceived by the threat radar. If the

Decoy

Figure 10.10 Initially, the threat radar centers its resolution cell on the target. The seduction decoy turns on within the threat resolution cell, presenting an RCS significantly larger than that of the target.

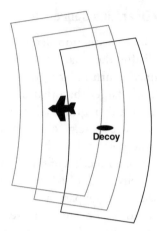

Figure 10.11   The greater RCS of the decoy causes the threat radar's resolution cell to
track it as it moves away from the target.

threat measures any signal return parameter that the decoy does not produce,
it will ignore the decoy and continue to track the target. Examples of parameters that might be important are jet engine modulation and effects related to
the size and shape of the target.

Figure 10.13 is a simplified representation of the RCS observed by the
threat radar during the decoy operating sequence. This drawing ignores the
geometrical effects of changing target orientation relative to the radar and
changing radar-to-target range. Both of these issues will be dealt with in the
next section.

Figure 10.12   When the threat radar's resolution cell is pulled far enough away so the target is no longer in the cell, the radar sees and tracks only the decoy.

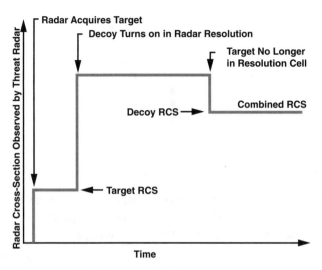

Figure 10.13 The larger RCS of the seduction decoy captures the tracking gates of the radar from the target.

10.6.2 Seduction Function in Ship Protection

Chaff bursts are used as seduction decoys for ship protection against radar-guided antiship missiles. In this case, the separation of the decoy from the target is generated only by the movement of the ship and by the wind, which moves the chaff burst. As shown in Figure 10.14, the chaff burst is ideally placed in the corner of the resolution cell from which it will separate from the ship most rapidly. The burst placement is chosen based on the type of radar in the attacking missile, the relative wind direction and velocity, and the direction from which the attack is coming. Seduction chaff bursts are often placed so close to the protected ship that chaff will fall on the deck of the ship.

10.6.3 Dump-Mode Decoy Operation

Another mode of decoy operation is important to ship defense. This is called "dump mode." In this operating mode, a decoy (for example, a chaff burst) is placed outside the radar's resolution cell, as shown in Figure 10.15. Then a deceptive jammer (for example, a range gate pull-off jammer) is used to pull the resolution cell away from the target and onto the decoy, as shown in Figure 10.16. As long as the decoy produces an RCS comparable to (and indistinguishable from) that of the protected target, the radar will lock onto the decoy. Naturally, the decoy must be placed in a position that will prevent the attacking missile from accidentally reacquiring the ship.

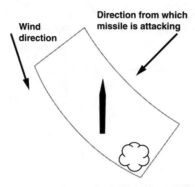

Figure 10.14 A chaff burst used in the seduction mode is placed in the corner of the resolution cell, which will cause the chaff burst to separate from the ship most rapidly in a direction that will lead the attacking missile away from the ship.

Figure 10.15 When ship-protection chaff is used in the dump mode, it is placed outside the threat radar's resolution cell.

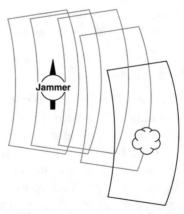

Figure 10.16 A deceptive jammer on the protected ship pulls the threat radar's resolution cell onto the chaff burst.

10.7 Effective RCS Through an Engagement

The effectiveness of a decoy is greatly affected by the circumstances in which it operates. Since virtually all decoy applications involve dynamic situations, it is very useful to consider what happens to a decoy through various engagement scenarios. Since the basically two-dimensional engagement scenarios involving the protection of ships against antiship missile attack are simpler to address, we will use these examples. However, the same principles will apply to aircraft protection when the appropriate engagement geometry is considered.

10.7.1 First, a Quick Review

The one-way link equation (discussed in Chapter 2) defines the received signal strength as a function of transmitter effective radiated power, signal frequency, and the distance between the transmitter location and the receiver location. Ignoring atmospheric losses, the equation is:

$$P_R = \text{ERP} - 32 - 20 \log (F) - 20 \log (d) + G_R$$

where P_R = signal power (in dBm); ERP = the effective radiated power of the transmitter (in dBm); F = the frequency of the transmitted signal (in MHz); d = the distance from the transmitter to the receiver (in km); and G_R = the receiving antenna gain (in dB).

As shown in Section 10.2, the equation for the effective gain (relative to isotropic antennas transmitting to and receiving from the decoy) as a function of RCS and signal frequency is:

$$G = -39 + 10 \log (\sigma) + 20 \log (F)$$

where G = the equivalent signal gain from the effective RCS of the decoy (in dB); σ = the effective RCS of the decoy (in m²); and F = the signal frequency (in MHz).

The $10 \log (\sigma)$ term is the RCS in dB relative to 1 m², or dBsm, and the equation can be reorganized into the form:

$$\text{RCS (dBsm)} = 39 + G - 20 \log (F)$$

10.7.2 A Simple Scenario

An antiship missile is fired against a ship from an aircraft and turns on its active tracking radar at the horizon (about 10 km from the ship). The ship,

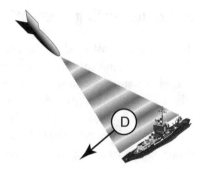

Figure 10.17   The engagement between the missile's active tracking radar, the decoy (D), and the target begins when the radar turns on. Both the decoy and the target will fall within the radar's antenna beam.

having been warned by its ESM system that an attack is imminent, has placed a decoy between itself and the missile. Both the decoy and the ship fall within the missile's radar beam, as shown in Figure 10.17. Assuming that the decoy is successful in capturing the missile's radar as the decoy moves away from the ship's location, the radar beam follows the decoy and the ship falls outside the radar beam, as shown in Figure 10.18.

Now, consider what the engagement looks like to the ESM receiver on the protected ship. If there were no decoy (or other EW) protection, the missile would move directly toward the ship at (typically) just under Mach 1, and the ship would remain at or near the center of the missile's radar beam. The signal power received by the ESM system would then have the time history shown in Figure 10.19. The ERP of the radar is the sum (in dB) of its transmitter power and its peak antenna gain. The ESM system's antenna gain remains constant, and the frequency remains constant. However, the signal propagation distance is reducing at the missile's closing speed, causing the

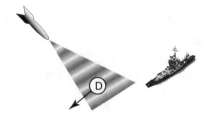

Figure 10.18   If the decoy captures the radar, the radar's antenna beam will be pulled away from the target as the missile homes on the decoy and the decoy moves away from the target.

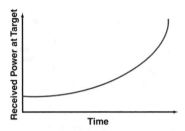

Figure 10.19 As the engagement proceeds, the received power at the target is the radar's ERP reduced by the (decreasing) propagation loss.

20 log (d) term to change rapidly. This term changes the propagation loss as a function of the square of the distance, so the received signal power will follow the curve shown in Figure 10.19.

Fortunately, the decoy captures the radar and moves its antenna beam away from the ship. As the ship leaves the radar antenna's main beam, the ERP of the radar in the direction of the ship drops off sharply, as shown in Figure 10.20. Incidentally, this is the same signal history that the decoy would see if it had not successfully captured the missile's radar—and it, rather than the ship, moved out of the antenna beam.

10.7.3 Decoy RCS Through the Scenario

The effective RCS of an active decoy depends on its gain and its maximum output power. As shown in Figure 10.21, the RCS (in dBsm) produced by a constant-gain decoy is 39 + Gain (dB) − 20 log (F) in MHz, until the missile gets close enough that the signal received by the decoy is equal to the

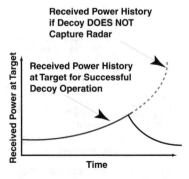

Figure 10.20 If the decoy captures the missile's radar, the signal power at the target drops as the target leaves the radar's antenna beam.

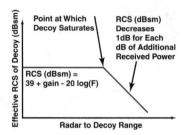

Figure 10.21 The effective radar cross-section of a decoy depends on its gain and its maximum output power.

maximum output power less the decoy gain. After that point, the effective RCS is reduced by 1 dB for each dB that the received signal power increases.

A primed oscillator decoy transmits at full power, regardless of the received signal power, so when it receives a very weak radar signal (i.e., at long range), the difference between the received and transmitted signals is very large. The decoy has, in effect, an immense gain and thus produces an immense effective RCS.

Figure 10.22 puts some numbers on the engagement. Consider an active decoy with 80-dB gain and a maximum output power of 100W, operating against a 10-GHz radar with a 100-kW ERP (nice round numbers; any similarity to any equipment, living or dead, is purely coincidental). The dashed line curve in the figure shows the effective RCS of the decoy as a function of its distance from the radar. In the linear gain range, the decoy produces an RCS (dBsm) of:

$$39 + G - 20 \log (10{,}000) = 39 + 80 - 80 = 39 \text{ dBsm}$$

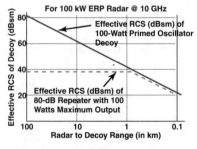

Figure 10.22 The effective RCS of a primed oscillator decoy varies inversely with the radar-to-decoy range. For a fixed-gain decoy, the RCS remains constant until the decoy saturates.

The decoy's RCS starts to roll off when the received signal from the radar is 100W − 80 dB (+50 dBm − 80 dB = −30 dBm). This occurs when the ERP − 32 − 20 log (F) − 20 log (d) = −30 dBm (assuming 0-dB receiving antenna gain). Plugging in some numbers and rearranging the equation yield:

$$20 \log (d) = 30 \text{ dBm} + \text{ERP} - 32 - 20 \log (F) =$$
$$30 + 80 - 32 - 80 = -2 \text{ dB}$$

where $d = 10^{(-2/20)} = 0.794$ km, or 794m.

If the decoy were a primed oscillator, always operating at its maximum power and with sufficient sensitivity to detect the signal at any appropriate range, the effective RCS would match the solid line in the figure. To calculate the effective RCS at 10 km, calculate the received signal strength at that range:

$$P_R \text{ (dBm)} = +80 \text{ (dBm)} - 32 - 20 \log (10) - 80 =$$
$$80 - 32 - 20 - 80 = -52 \text{ dBm}$$

Since the decoy outputs 100 W, the effective gain is +50 dBm − (−52 dBm) = 102 dB. This makes the effective RCS (dBsm):

$$\text{RCS(dBsm)} = 39 + G - 20 \log (F) = 39 + 102 - 80 = 61 \text{ dBsm}$$

This is more than 1 million m^2.

11

Simulation

Generally, EW simulation is employed to save money. However, there are other perhaps even more pressing reasons to simulate something. Simulation can allow the realistic evaluation of the performance of operators, equipment, and techniques under circumstances that don't yet exist. Simulation can also allow the realistic training of individuals under conditions that, in real life, might kill them.

11.1 Definitions

Simulation is the creation of an artificial situation or stimulus that causes an outcome to occur as though a corresponding real situation or stimulus were present. EW simulation often involves the creation of signals similar to those generated by an enemy's electronic assets. These artificial signals are used to train operators, to evaluate the performance of EW systems and subsystems, and to predict the performance of enemy electronic assets or the weapons they control.

Through simulation, operators and EW equipment can be caused to react as though one or more threat signals were present and doing what they would during a military encounter. Usually, the simulation involves interactive updating of the simulated threat as a function of operator or equipment responses to the threat signals.

11.1.1 Simulation Approaches

Simulation is often divided into three subcategories: computer simulation, operator interface simulation, and emulation. Computer simulation is also called "modeling." Operator interface simulation is often simply called "simulation." This can cause confusion because the same term is commonly used to define both the whole field, and this particular approach. All three approaches are employed for either training or test and evaluation (T&E). Table 11.1 shows the frequency of the use of each subcategory for each purpose.

11.1.2 Modeling

Computer simulation (or modeling) is done in a computer using mathematical representations of friendly and enemy assets and evaluating how they interact with each other. In modeling, no signals or representations of tactical operator controls or displays are generated. The purpose is to evaluate the interaction of equipment and tactics that can be mathematically defined. Modeling is useful for the evaluation of strategies and tactics. A situation is defined, each of several approaches is implemented, and the outcomes are compared. It is important to note that any simulation or emulation must be based on a model of the interaction between an EW system and a threat environment as shown in Figure 11.1.

11.1.3 Simulation

Operator interface simulation refers to the generation of operator displays and the reading of operator controls in response to a situation that has been modeled and is proceeding—without the generation of actual signals. The operator sees computer-generated displays and hears computer-generated audio as though he or she were in a tactical situation. The computer reads the operator's control responses and modifies the displayed information accordingly. If the operator's control actions could modify the tactical situation, this is also reflected in the display presentations.

Table 11.1
Simulation Approaches and Purposes

Simulation Purpose	Simulation Approach		
	Modeling	**Simulation**	**Emulation**
Training	Commonly	Commonly	Sometimes
Test & evaluation	Sometimes	Seldom	Commonly

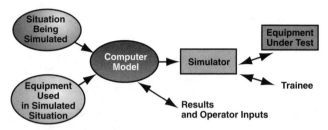

Figure 11.1 Any type of simluation must be based on a model of equipment and/or a tactical situation.

In some applications, operator interface simulation is achieved by driving the system displays from a simulation computer. Switches are read as binary inputs, and analog controls (a rotated volume control, for example) are usually attached to shaft encoders to provide a computer-readable knob position.

The other approach is to generate artificial representations of the system displays on computer screens. Displays are represented as pictures of the system displays, usually including a portion of the instrument panel in addition to the actual dials or CRT screens. Controls are depicted on the computer screen and operated by a mouse or touch-screen feature.

11.1.4 Emulation

Where any part of the actual system is present the emulation approach is used. Emulation involves generation of signals in the form they would have at the point where they are injected into the system. Although the emulation approach *can* be used for training, it must *almost always* be used for the T&E of systems or subsystems.

As shown in Figure 11.2, emulated signals can be injected at many points in the system. The trick is to make the injected signal look and act like it has come through the whole system—in the simulated tactical situation. Another important point is that anything that happens downstream from the injection point may have an impact on the signal arriving at the injection point. If so, the injected signal must be appropriately modified.

11.1.5 Simulation for Training

Simulation for training exposes students to experiences (in a safe and controlled way) that allow them to learn or practice skills. In EW training, this most often involves experiencing enemy signals in the way they would be

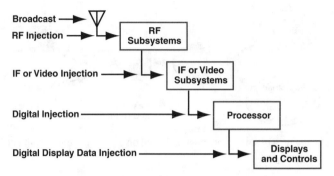

Figure 11.2   Emulated signals can be injected at many points in an EW system, usually to provide realistic testing of the subsystems immediately downstream of the injection point.

encountered if the trainee were at an operating position in a military situation. EW simulation is often combined with other types of simulation to provide a full training experience. For example, a cockpit simulator for a particular aircraft may include EW displays which react as though the aircraft were flying through a hostile electronic environment. Training simulation usually allows an instructor to observe what the student sees and how he or she responds. Sometimes, the instructor can play back the situation and responses as part of the debriefing after a training exercise—a powerful learning experience.

11.1.6   Simulation for T&E

Simulation for equipment T&E involves making equipment think it is doing the job for which it was designed. This can be as simple as generating a signal with the characteristics a sensor is designed to detect. It can be as complex as generating a realistic signal environment containing all of the signals a full system will experience as it moves through a lengthy engagement scenario. Further, that environment may vary in response to a preprogrammed or operator-selected sequence of control and movement actions performed by the system being tested. It is distinguished from training simulation in that its purpose is to determine how well the equipment works, rather than to impart skills to operators.

11.1.7   Fidelity in EW Simulation

Fidelity is an important consideration in the design or selection of an EW simulator. The fidelity of the model and of the data presented to systems and

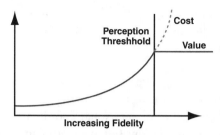

Figure 11.3 The cost of simulation can increase exponentially, while its value does not increase beyond the threshold at which the equipment under test or the individual being trained can no longer perceive inaccuracies.

operators must be adequate to the task. In training simulation, the fidelity must be adequate to prevent the operator from detecting the simulation (or at least to avoid interfering with the training objectives). In T&E simulation, the fidelity must be adequate to provide injected signal accuracy better than the perception threshold of the tested equipment. As shown in Figure 11.3, the cost of a simulation rises—often exponentially—as a function of the fidelity provided. However, the value of the fidelity does not increase once the perception level of the trainee or tested equipment is reached.

11.2 Computer Simulation

Computer simulation involves setting up a model of some situation or equipment and manipulating that model to determine some outcome. Important simulations in the EW field include:

- Analysis of the performance of an EW asset against a threat scenario in which signals from one or more threat emitters are applied in response to the sequence expected in some combat situation;

- Engagement analysis of an electronically controlled weapon and its target, including the effects of applying various EW assets;

- Analysis of the survivability of a friendly aircraft, ship, or mobile ground asset protected by various EW capabilities as it moves through a typical mission scenario.

11.2.1 The Model

A computer simulation is based on a model in which all pertinent characteristics of every "player" are expressed mathematically. There is a "gaming area" in which the players interact. Note that the gaming area can have many dimensions, such as location, frequency, time, and so on. The steps in construction of the model are:

- Design the gaming area. How much area does the action cover? Include the elevation of the highest player. How far away can a player affect another player? What coordinate system is most convenient for the simulation? Often a Cartesian coordinate system can be used (x and y along a flat zero-altitude surface and z for elevation) with zero being at one corner of the gaming area.
- Add terrain elevations to the gaming area if appropriate.
- Characterize each player. What are its qualities? What specific actions by other players cause changes in those qualities? How does it move? Each of these qualities must be described numerically, and the movements must be described in terms of equations with the specific actions as elements of the equations. Determine the required model resolution and set the model timing increments.
- Set the players in their initial positions and conditions.

Once the model is set up, the simulation can be run and the results determined.

11.2.2   Ship-Protection Model Example

An example of an EW engagement model is shown in Figure 11.4. This models an engagement between a ship and a radar guided antiship missile. The ship is protected by a chaff cloud and a decoy. The simulation determines the distance by which the missile misses the ship. If the miss distance is less than the size of the ship, the missile wins.

The gaming area must be large enough to contain all of the action. The missile will turn on its radar as it reaches the ship's radar horizon (about 10 km away). Since we don't know the direction of attack, the gaming area must contain at least a 10-km circle around the ship. Unless the terminal movement of the missile (which may climb and dive) is included in the simulation, the gaming area can be two-dimensional. The only other element of the gaming area is the wind, which has speed and direction.

The players include the ship, missile, decoy, and chaff burst.

The ship has location, velocity vector, and radar cross-section (RCS). If the ship is not turning, its location can be calculated by moving it from its

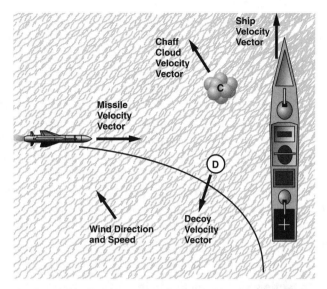

Figure 11.4   A model for the engagement between a missile and an EW-protected ship will also include all of the protecting assets.

starting location in its direction of travel at its steaming speed. In certain conditions, it is appropriate to make a maximum-rate turn when a missile's radar is detected. For any type of ship at any steaming rate, the forward and cross-path locations for a maximum-rate turn will be available as a table. The path describes a spiral, since the turn causes the ship to slow. The ship's RCS will be available in graphic or tabular form as a function of angle from the ship's bow and elevation angle.

The missile has location, velocity vector, and radar parameters. Normally, the missile flies at a constant speed, at a small, fixed distance above the surface of the ocean. Its direction of travel will be determined by what its radar receives. The missile's location differs from its last location by its speed times the calculation interval—in the direction dictated by its radar. Determining that direction is the essence of the simulation and will be covered below. We will take the missile's radar to be a pulse type with a vertical-fan beam. The significant radar parameters include effective radiated power (ERP), frequency, pulse width, horizontal-beamwidth, and scan parameters. If the ship is detecting the incoming missile by radar, the missile's RCS is also significant. For this example, we will assume that the ship only detects the missile's radar.

The chaff cloud has location, velocity vector, and RCS. Its velocity vector is determined by the wind, since the cloud drifts with the wind. The RCS of the chaff cloud is fixed for any specific frequency (i.e., the operating frequency of the missile radar). For this simulation, it is assumed that the chaff cloud is at the optimum altitude to maximize the RCS.

The decoy has location, velocity vector, gain, and maximum output power. It receives signals from the radar and retransmits them at the maximum possible ERP—that is, the received power increased by the decoy's throughput gain (including antenna gains). The throughput gain of the decoy creates an effective RCS. It is interesting to note that a primed-oscillator decoy, which transmits at its maximum power at the frequency of the received signal, will create maximum RCS when the received radar signal is at its minimum (i.e., at maximum range). The RCS would then reduce as the square of range as the missile approaches the decoy. If the decoy floats, it will have no movement at all. It could also be a type of decoy that moves through some preset pattern to seduce a missile away from the ship.

11.2.3 Ship-Protection Simulation Example

The simulation begins with the ship steaming at some azimuth and the missile turning on its radar 10 km from the ship at some azimuth while traveling toward the ship at its cruise speed.

The missile radar sweeps through an angular segment until it acquires a target. If the decoy or the chaff cloud have been deployed before the radar turns on, the missile may acquire one of them instead of the ship; however, we will assume the worst case—that the missile radar acquires the ship. After acquisition, the radar steers the missile toward the ship's radar return.

As shown in Figure 11.5, the missile radar has a resolution cell of depth equal to the pulse width times 0.3m/nanosecond. (Note that this value is commonly used for calculations involving ship protection with chaff, while half this value is used for airborne or land-based radars.) The width of the

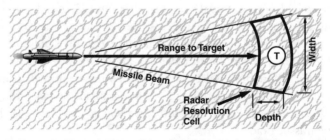

Figure 11.5 The radar's resolution cell is a function of pulse width and antenna-beam width.

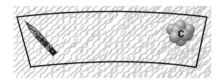

Figure 11.6 The missile's radar keeps its resolution cell centered on the location from which the maximum return signal seems to be originating. That can be the combination of two or more objects within its resolution cell.

resolution cell is twice the sine of half the radar's 3-dB horizontal beamwidth times the range from the radar to the target. The resolution cell is the area in which the radar cannot discriminate between two target returns. If there are two targets in the cell, the radar will respond as though there were one target located between the two, closer to the position of the stronger return, proportional to their relative RCS (see Figure 11.6). When the missile is far from the ship, the resolution cell is wide, and as it gets closer, the cell narrows. If the missile hits the ship, the cell would have a width of zero at the moment of impact.

As shown in Figure 11.7, during each calculation interval, the missile moves a distance equal to its speed times the calculation interval in the direction toward the apparent target location.

Considering the missile radar's point of view, the RCS of the ship is its RCS at the radar's frequency and at the angle of the radar from the bow of the ship. If the ship is turning, or if the aspect angle of the missile changes, the RCS changes. The RCS of the chaff cloud remains constant throughout the engagement. The RCS of the decoy is a function of the received power from the radar and the decoy's ERP.

As the resolution cell narrows, either the protecting device or the ship will fall out of the resolution cell—assuming that they are not lined up in the direction of the missile. If the protecting device provides a stronger RCS at

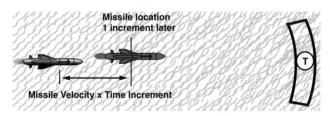

Figure 11.7 During one time increment, the missile moves a calculated distance toward the middle of its radar-resolution cell.

the critical time, it will capture the missile's resolution cell and protect the ship. If the separation geometry or the relative RCS is not adequate to capture the resolution cell, the missile will hit the ship.

11.3 Engagement Scenario Model

In Section 11.2 we discussed a computer model of an engagement between an antiship missile and a ship protected by chaff and decoys. In this section, we are putting some numbers into a simplified model to show how the model of the engagement would be implemented. The object of the analysis is to determine if the missile misses the ship and, if so, by how much. Although any type of computation program could be employed, we will use a spreadsheet.

Please remember that it is not suggested that the tactics used are the best way to defend the ship. The purpose is to use the model to determine what happens if these tactics *are* used. Figure 11.8 shows the situation simulated. It includes several simplifications for reasons of length: The radar's resolution cell is represented as a rectangle; the ship's RCS profile is unrealistically simple; we consider only the missile, the ship, and the chaff cloud; and we start the engagement with the chaff cloud already bloomed to its full RCS. All values are input to the model in consistent units.

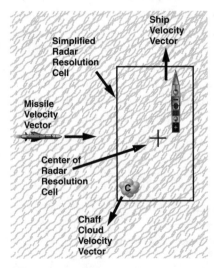

Figure 11.8 In this simplified engagement model, the ship is protected only by chaff; the missile flies toward the center of the radar resolution cell.

11.3.1 Numerical Values in the Model

The following numerical values are assigned to the descriptions of the gaming area and the players in the engagement: The gaming area is open ocean with a 2.83 meter/sec.-wind from azimuth 45°. The gaming area has a two-dimensional coordinate system centered on the ship. The engagement has 1-sec time resolution. The ship is traveling due north at 12 m/sec and retains this bearing throughout the engagement. The radar RCS of the ship is described by Figure 11.9.

We start the simulation with the missile at an azimuth of 270°, 6 km from the ship. Its radar is locked onto the ship. The missile travels at 250 m/sec close to the surface of the water and has a radar with a 5°-wide vertical-fan beam antenna. We assume that the antenna has the same gain across its beam and zero gain outside its beam. The radar's pulse width is 1 μsec. The missile steers toward the center of its resolution cell, which is centered on the apparent radar return within the cell (i.e., if there are two items in the cell, it centers between them, proportionally closer to the larger RCS). The chaff cloud—bloomed to its full RCS of 30,000 m²—is located in the lower left corner of the radar's resolution cell. Since decoys, jammers, or ESM systems are not included in the problem, and since both the ship and the chaff cloud are passive radar reflectors, it is not necessary to specify the radar's effective radiated power, antenna gain, or operating frequency.

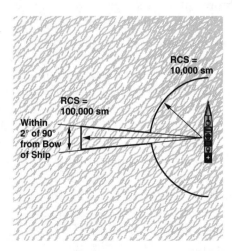

Figure 11.9 For this analysis, the RCS of the protected ship is 10,000 m², except within 2° of 90° from the bow of the ship. The RCS is symmetrical (right and left).

11.3.2 Ship Defense with Chaff

The optimum placement for the chaff is within the radar's resolution cell in the direction that will cause the wind to create the maximum separation between the chaff cloud and the ship.

The resolution cell of the radar is centered on the ship until the chaff cloud enters the resolution cell—then the cloud and the ship separate. The chaff drifts downwind while the ship steams away. Our simulation is to determine the relative positions of the missile and the ship through the engagement.

First, consider the initial locations of the players in the gaming area: The ship is at the origin (0/0); the chaff cloud is at $x = -125m$, $y = -250m$ ($-125/-250$); and the missile is at ($-6000/1$). The chaff's velocity vector is 2.83 m/sec at azimuth 225°, because it drifts with the wind. The missile's velocity vector is 250 m/sec in the direction of the apparent target in its resolution cell (i.e., the center of the resolution cell as shown in Figure 11.10). The radar sees all targets in its resolution cell and adjusts the location of its cell to the apparent location of the sum of the RCS of all objects in the resolution cell.

Each second, the engagement scenario program calculates the location and velocity vectors of all players using formulas on a spreadsheet. Figure 11.11 is a diagram used to calculate whether or not the ship and/or the chaff cloud remain in the radar's resolution cell at each calculation point.

Figure 11.12 is a spreadsheet showing how to set up the calculations. Rows 2–16 contain the problem's input parameters. Rows 19–42 are the actual calculations for the engagement. Column B shows the condition of the engagement when it begins. Column C shows the locations of all players after one second. It also determines the velocity vectors and determines whether or not the ship or the chaff cloud are still in the resolution cell. Column D shows the formulas which are entered into column C of the spread sheet. You will note (perhaps with some alarm) that only the first second of the engagement is shown. If you want to find out if the missile hit the ship, you will have to enter the formulas into column C, get rid of column D, and copy column C into many subsequent columns. The formulas automatically increment to the correct reference cells. Row 31 shows the missile-to-ship distance. If this distance ever becomes zero, the ship is hit. If the missile misses the ship, the missile-to-ship distance will pass through a minimum (the miss distance) and increase again.

As you run the problem, you will note that the RCS of the ship dominates the early part of the engagement until the geometry changes to move the missile out of the ship's high RCS aspect angle. The formula in row 24

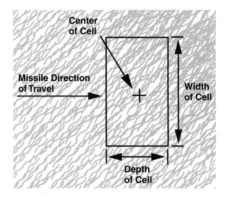

Figure 11.10 For this example, we use a simplified rectangular representation of the radar's resolution cell.

calculates an azimuth with an arc-tangent formula. The complexity of the formula is required by the nature of the function. In the calculations to determine whether the ship or the chaff cloud are within the missile radar's resolution cell, the values shown in Figure 11.11 are derived using trigonometric identities to avoid the complication of calculating any actual angles.

Also, when the missile gets within 1 sec of the ship, you will probably want to increase the time resolution to 0.1 sec to get a good reading on the minimum missile-to-ship distance.

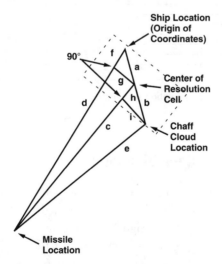

Figure 11.11 This figure is used to calculate the values to determine whether the ship and/or the chaff cloud are within the radar's resolution cell.

	Column A	Column B	Column C	Column D
1	Initial Conditions			
2	Ship travel (azimuth)	0		
3	Ship speed (m/sec)	12		
4	Missile x value (m)	-6000		
5	Missile y value (m)	1		
6	Missile azimuth (deg)	90		
7	Missile speed (m/sec)	250		
8	Radar frequency (GHz)	6		
9	Radar PW (usec)	1		
10	Radar beam width (deg)	5		
11	Chaff cloud x value	-125		
12	Chaff cloud y value	-250		
13	Chaff cloud RCS	30000		
14	Wind direction (azimuth)	225		
15	Wind speed (m/sec)	2.83		
16	Ship RCS to missile (sm)	100000		
17				
18	Engagement calculations			Formulas
19	Time (sec)	0	1	
20	Missile x value	-6000	-5750	=B20+$B$7*SIN(B24/57.296)
21	Missile y value	1	1.001511188	=B21+$B$7*COS(B24/57.296)
22	Ship in cell? (1=yes)	1	1	=IF(AND(C39<(C33/2),C40<(C32/2)),1,0"
23	Chaff in cell? (1=yes)	1	1	=IF(AND(C41<(C33/2),C42<(C32/2)),1,0"
24	Missile vector azimuth	90	90.59210989	=IF((C27-C20)>0,IF((C28-C21)>0,ATAN ((C27-C20)/(C28-C21))*57.296,180+ATAN ((C27-C20)/(C28-C21))*57.296),IF((C28-B21) <0,ATAN((C27-C20)/(C28-C21))*57.296+180, 360+ATAN((C27-C20)/(C28-C21))*57.296))"
25	Chaff x value	-125	-127.0010819	=$B$11+$B$15*SIN($B$14/57.296)*C19
26	Chaff y value	-250	-252.0011424	=$B$12+$B$15*COS($B$14/57.296)*C19
27	Center radar cell x value	-96.15384615	-29.30794199	=C25*($B$13*C23/($B$13*C23+C30*C22))
28	Center radar cell y value	-192.3076923	-58.15410979	=C26*($B$13*C23/($B$13*C23+C30*C22))
29	Bow-to-radar angle (deg)	90	90	=ABS(180-B24)
30	Ship RCS (to missile)	100000	100000	=IF(B29<88,10000,IF(B29<92,100000, 10000))"
31	Missile-to-ship distance	6000	5750.000087	=SQRT(C20²+C21²)
32	Radar cell width	523	515.3183339	=2*SIN($B$10/(2*57.296))*SQRT((B27-B20)² +(B28-B21)²)
33	Radar cell depth	305	305	=$B$9*305
34	a in **Figure 11.11**		65.12185462	=SQRT(C27²+C28²)
35	b in **Figure 11.11**		282.1947034	=SQRT(C25^2+C26²)-B34
36	c in **Figure 11.11**		5720.997903	=SQRT((C27-C20)^2+(C28-C21)²)
37	d in **Figure 11.11**		5750.000087	=C31
38	e in **Figure 11.11**		5628.687873	=SQRT((C25-C20)²+(C26-C21)²)
39	f in **Figure 11.11**		29.2978125	=(C36²-C34²-C37²)/(-2*C37)
40	g in **Figure 11.11**		58.15921365	=SQRT(C34²-C39²)
41	h in **Figure 11.11**		98.52509165	=(C38²-C36²-C35²)/(-2*C36)
42	i in **Figure 11.11**		264.4364894	=SQRT(C35²-C41²)

Figure 11.12   Engagement calculation spreadsheet.

11.4   Operator Interface Simulation

An important class of simulation reproduces only the operator interface. This is sometimes called just "simulation" as opposed to "emulation," which involves the generation of actual signals at some point in the process to drive the operator interface. In operator interface simulation, it is only the part of the process that the operator sees, hears, and touches that is included. Everything that takes place behind the scenes is transparent to the operator—and thus is important only as it is reflected in the operator interfaces.

In most situations, it is practical to simulate a military engagement or equipment interaction of some kind all in software. It can then be determined what the operator would see, hear, and feel if he or she were in this postulated situation. It is also practical to sense what actions the operator takes and determine how the situation would be changed in response to those actions—and how that change would be sensed by the operator. The operator interface simulator typically works from a digital model of the equipment and the engagement to determine the appropriate operator interfaces and present those to the operator.

If the operator actions are sensed in real time (with adequate fidelity) and the resulting situation is experienced by the operator in real time (with adequate fidelity), the operator will have the necessary training experience.

So far, we could well have been talking about a flight simulator, but all of this applies equally well to simulation that teaches the operation of EW equipment. In fact, it could be (and is) applied to training in the operational use of EW equipment in the situation being simulated in the flight simulator—or any other military platform.

11.4.1 Primarily for Training

First, understand that the simulation of the operator interface has no role in the evaluation of how well the equipment being simulated does its job. Its principal value is to train the operator in a range of tasks that vary from simple "knobology" (i.e., what knob does what?) to the sophisticated use of EW equipment in extremely stressful situations, to give realistic combat experience with no casualties. A secondary use of operator interface simulation is to evaluate the adequacy of the operator interface provided by a system, determining if the system's controls and displays are adequate to allow the operator to do the job that must be done.

11.4.2 Two Basic Approaches

There are two basic approaches to the simulation of operator interfaces. One is to provide the actual control and display panels from the system, but to drive them directly from a computer, as shown in Figure 11.13. This approach has the advantage of a realistic operator-training experience. The real knobs are where they should be—and are the right size and shape. The displays have the right level of flicker and so forth. There are three problems with this approach: The equipment may be expensive "mil-spec" hardware, it needs to be maintained, and it typically requires special hardware and software to interface the system hardware to the computer. Mil-spec hardware is

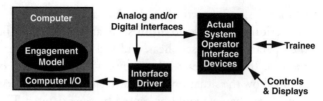

Figure 11.13   Operator interface simulation can be implemented with actual system-control and displays panels, driven directly by a simulation computer.

expensive, and the extra interface equipment must be built and maintained. All of these considerations add cost.

The displays are driven by injecting display signals in the form and format they would have if the hardware were being used operationally. Likewise, the signals coming from the operator control/display panels are sensed and converted to a form most convenient for the computer to accept.

The second approach is to use standard, commercial computer displays to simulate the operational displays. The controls can be created from commercial parts, or they can be simulated on the computer screen and accessed by keyboard or mouse as in Figure 11.14.

Figure 11.15 shows a simulation of an AN/APR-39A radar-warning-receiver cockpit display on a computer screen. The symbols on the screen move in response to simulated aircraft maneuvers. In this simulation, the control switches are also displayed on the screen. If the operator clicks on a switch with the mouse, the switch changes position on the screen and the system response to the switch action is simulated.

If a simulated control panel is used, rather than pictures of controls on the computer screen, the controls need to be sensed and their positions entered into the computer. Figure 11.16 shows the basic technique. Each switch provides logic level "one" voltage to a specific location in a digital register when the switch is on. The exact voltage depends on the type of logic used. Alternately, the switch may just ground that location when it is on. For an analog control (e.g., a knob), a shaft encoder is used. The shaft

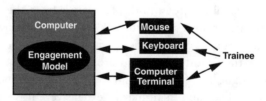

Figure 11.14   The operational system's controls and displays can be represented by use of standard commercial computer peripheral devices.

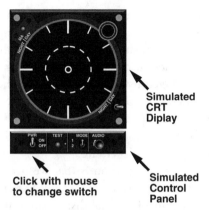

Simulated
CRT
Diplay

Click with mouse
to change switch

Simulated
Control
Panel

Figure 11.15 This is a reproduction of a computer simulated operator interface (original courtesy of I3C Inc.).

encoder typically provides pulses every few degrees as the control is moved, and the up/down counter converts those pulses into a digital-control position word which is input to the proper location in the register.

The register is periodically read by the computer to sense the control positions. This is a very low-rate process because of the low rate at which our hands move.

11.4.3 Fidelity

The fidelity necessary in a simulated operator interface is determined by simple criteria. If the operator can't perceive it, it need not be included in the simulation. The elements of fidelity are control-response accuracy, display accuracy, and the timing accuracy of both. First, let's deal with the time

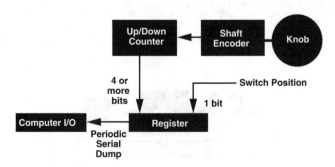

Figure 11.16 In a simulated control panel, the control positions must be converted into digital words for input to the computer.

fidelity. The human eye requires about 42 ms to take in an image. Thus, if a display is updated 24 times per second (as with motion pictures), the operator perceives smooth action. In simulations that bring the operator's peripheral vision into play, the action must be quicker. Your peripheral vision is faster, so you will be bothered by flicker in your peripheral vision from the frames in a 24 frame/second presentation. In movies, which may be shown on wide-angle screens, the approach is to have 24 frames/second—but to flash the light twice for each frame so your peripheral vision will not be able to follow the 48 frames/second flicker rate.

Another perceptual consideration is that we perceive changes in patterns of light and dark (i.e., motion) more rapidly than we perceive color changes. These two elements of a visual display are called "luminance" and "chrominance." In video compression schemes, it is common practice to update the luminance at twice the chrominance update rate.

A time-related consideration critical to simulation is the perception of the results of actions we have taken. Magicians say, "The hand is quicker than the eye," but this is simply not true. Even the quickest hand motion (for example, pushing a stop-watch button a second time) takes 150 ms or more. Try it to see how short a time you can catch on your digital watch. However, you can perceive visual changes more rapidly. For example, when you turn on a light switch, you expect the light to come on immediately. As long as it actually comes on within 42 ms, your experience will be the same as the real-world situation. (See Figure 11.17.) In operator interface simulation, the simulator must track binary-switch actions and analog-control actions such as the turning of a knob. While our perception of the results of a knob turning are less precise, it is still good practice to assume that the knob position should be translated into a visual response within 42 ms for full time fidelity.

Location accuracy is a little more tricky. We humans are not very good at perceiving absolute values of location or intensity—but we are very good at determining *relative* location or intensity. This means that if two items are supposed to be at the same angle or distance, we can sense a very small difference between their angles or distances. On the other hand, if both are off

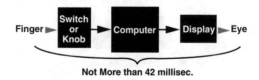

Figure 11.17   For ideal fidelity, the total time from a control action to the display representation should not exceed 42 milliseconds.

(together) by a few degrees or a few percent of range, we will probably not notice. This drives the requirement for indexing gaming areas, which will be covered later.

11.5 Practical Considerations in Operator Interface Simulation

Some less-than-obvious considerations must be satisfied in operator interface simulation. One is the coordination of EW simulation with other types of simulation. Another is the representation of anomalous effects in real hardware. A third is process latency. Finally, there is the concept of "good enough."

11.5.1 Gaming Areas

We talked briefly about gaming areas in Section 11.2 as being big enough to hold all of the "players" in the simulation. For clarity, consider what the gaming area means in the model that supports a flight simulator with EW incorporated.

As shown in Figure 11.18, the gaming area is a box of space above the ground. Its altitude is the maximum operating altitude of the highest aircraft

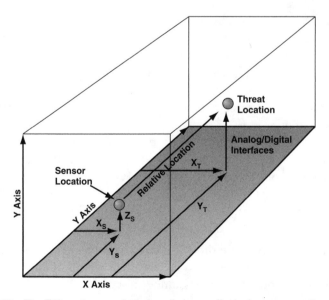

Figure 11.18   The EW gaming area includes all players. Each player senses the other players relatively and according to its own sensor perception.

(or threat aircraft) simulated. Its x and y axes cover the whole flight path of the simulated aircraft mission and all of the threats (air or ground) that will be observed by systems in the simulated aircraft.

Any sensor mounted on the simulated aircraft will have the same location in the gaming area as the aircraft. Its x, y, and z values are determined by flight-control manipulations of the simulator's "pilot." The x, y, and z values for the threat are determined by the model.

The sensor's view of the threat is determined by their instantaneous, relative locations. The range and aspect angles are calculated from the x, y, and z values. The range and angles, in turn, determine what is presented on the cockpit displays in the simulator.

11.5.2 Gaming Area Indexing

There are usually several gaming areas in a flight simulator—one for each type of sensor in the aircraft. The EW gaming area includes all EW threats. The radar land-mass gaming area includes the shape of the terrain. The visual gaming area includes everything the pilot sees—terrain, buildings, other aircraft, and visual aspects of threats.

To provide realistic training in a flight simulator, it is important that the situation presented to the operator by the various cockpit displays be consistent. The location and apparent size of an enemy aircraft as seen from the cockpit will be determined by the range and the relative positions and orientations of the simulated aircraft and the modeled enemy aircraft. As shown in Figure 11.19, the EW system display (here a radar warning receiver screen) should indicate an appropriate threat symbol at the appropriate screen location for that type of enemy aircraft at that range and aspect angle. Likewise, the plan position indicator radar display shows a return at the appropriate range and azimuth.

The mechanism for controlling the consistency of the various cues given to the simulator pilot is to specify the indexing of the various gaming areas. If, for example, the specification were 100-ft indexing of gaming areas:

- The radar and the visual displays would need to show any point on the ground at the same elevation within 100 ft.
- The location of the simulated aircraft and all other "players" in the simulation would need to be consistent within 100 ft with equivalent locations in all other gaming areas.
- Line-of-sight determinations and the resulting changes in displayed data would be appropriate within 100 ft.

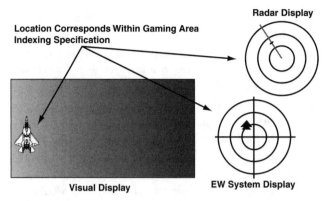

Figure 11.19 The indexing of gaming areas determines the degree of accuracy to which the information on the various system displays and the visual simulations are consistent.

11.5.3 Hardware Anomalies

One principle of simulation is that the simulator design is responsible for everything that takes place "up stream" of the simulation point. In Figure 11.20, the simulator provides its inputs at the "simulation point." It simulates all equipment and/or processes to the left of the simulation point—to be acted on by the equipment and processes to the right of the simulation point. Anything wrong with the hardware and processes to the right of the simulation point will have the same effect on the simulated inputs that they would have on real-world inputs. However, the anomalies in the left-hand processes (which are not really present because they are being simulated) will not be present unless they are included in the simulation.

It is easy to assume that all simulated hardware and software work as advertised, and to incorporate that proper operation into the simulation. Unfortunately, when equipment meets the real world, it sometimes becomes "creative." Since an operator interface simulator simulates most or all of the hardware, it is typically responsible for all equipment anomalies.

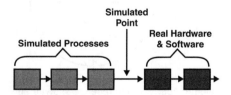

Figure 11.20 The simulator is responsible for everything upstream of the simulation point.

Figure 11.21 presents a real-world example of a hardware anomaly in an EW system. One early digital RWR had a system processor that collected threat-location data in "bins" (fixed memory locations). The system collected current threat data from each active location (range and angle of arrival relative to the aircraft) and would store each intercept for a fixed amount of time before "timing out" the old data and discarding it. The timing of data storage was such that the aircraft could turn fast enough to activate new bins before the data in the old bins had timed out. Thus, a single SA-2 missile site could look like several such sites at various angles under some operational conditions. The situation shown represents actual range data taken as the aircraft made a high-G turn to the left while receiving a strong SA-2 signal.

If that system had been simulated in a training simulator, and this hardware anomaly were not reproduced, the students training against the simulator would have been trained to expect only the correct symbol. This is called "negative training" and is to be avoided if at all possible.

11.5.4 Process Latency

A typical hardware/software problem that must be considered in operator interface simulator design is process latency. Latency is the time required for some process to be completed. Particularly in older systems with slower computers, there can be processing delays that are significant compared to the 24 frames/second update rate of the operator's eyes. This can, for example, cause symbols to be placed improperly when an aircraft is in a high-rate roll.

The processing required for a simulator to artificially create the data resulting from some tactical situation may be either more or less complex than the processing required for the system to do the job in the real world.

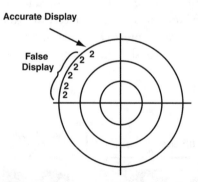

Figure 11.21 An early digital radar warning receiver showed additional false threat displays when the aircraft was in a high-G turn.

Also, the simulation computer can be either faster or slower than the real-world computer. It is important to assure that process latency in the simulator (either too much or too little) does not create negative training.

11.5.5 Yes, But Is It Good Enough?

We have been talking about the fidelity required in simulation in terms of what a human being can perceive. That is not always the right answer. Fidelity is often a significant cost driver in simulators, and more fidelity than required to do the training job is a waste of money. The real question is, "How good does the simulation have to be for the trainee to achieve the skills the training is designed to impart?" A good example is the quality of graphics in visual displays. If an enemy aircraft looks "chunky" but moves properly, the training will be effective. Note that these comments relate specifically to simulation for training; a different standard will apply when we talk about simulation for test and evaluation of equipment.

11.6 Emulation

Emulation involves the generation of real signals to be received by a receiving system—or by some part of that system. This is done either to test the system (or subsystem) or to train operators in the operation of the equipment.

In order to emulate a threat emission, it is necessary to understand all of the elements of the transmitted signal and to understand what happens to that signal at each stage of transmission, reception, and processing. Next, a signal is designed to look like the signal at some specific stage in the path. That signal is generated and injected into the process at the required point. The requirement is that all of the equipment downstream from the injection point "think" it is seeing the real signal in an operational situation.

11.6.1 Emulation Generation

As shown in Figure 11.22, emulation, like any other type of simulation, starts with a model of what must be simulated. First, of course, the characteristics of the threat signals must be modeled. Then, the way the EW system will experience those threats must be modeled. This engagement model determines which threats the system will see and the range and angles of arrival at which the system will see each threat. Finally, there must be some kind of model of the EW system. This system model (or at least partial system model) must exist, because the injected signal will be modified to simulate

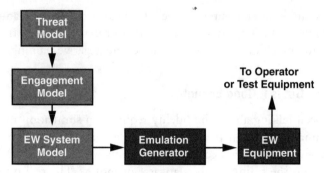

Figure 11.22 Emulation generates injection signals based on the model of the threats, the engagements, and the equipment into which the signals are injected.

the effects of all of the parts of the system that are upstream of the injection point. The upstream components may also be affected by actions of downstream components. Examples of this are automatic gain control and anticipated operator-control actions.

11.6.2 Emulation Injection Points

Figure 11.23 is a simplified diagram of the whole transmission, reception, and processing path for a threat signal, and the points at which emulated signals can be injected. Table 11.2 summarizes the simulation tasks required by the selection of each injection point, and the following discussion expands upon the associated applications and implications.

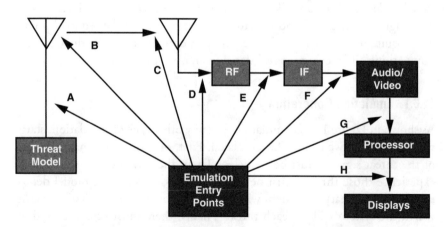

Figure 11.23 Emulations of threat signals can be injected at many different points in the transmission/reception/processing path.

Table 11.2
Emulation Injection Points

Injection Point	Injection Technique	Simulated Parts of the Path
A	Full capability threat simulator	The threat modulation and modes of operation
B	Broadcast simulator	Threat modulation and antenna scanning
C	Received signal energy simulator	Transmitted signal, transmission path losses, and angle-of-arrival effects
D	RF signal simulator	Transmitted signal, path losses, and receiving-antenna effects, including angle of arrival
E	IF signal simulator	Transmitted signal, path losses, receiving-antenna effects, and RF-equipment effects
F	Audio- or video-input simulator	Transmitted signal, path losses, receiving-antenna and RF-equipment effects, and effects from the selection of IF filters
G	Audio- or video-output simulator	Transmitted signal, path losses, receiving-antenna and RF-equipment effects, and effects from the selection of IF filters and demodulation techniques
H	Displayed signal simulator	The entire transmission/reception/processing path

Injection Point A. Full-Capability Threat Simulator: This technique creates an individual threat simulator that can usually do everything that the actual threat can. It is typically mounted on a carrier that can simulate the mobility of the actual threat. Since it uses real antennas like those of the threat emitter, antenna scanning is very realistic; multiple receivers receive a scanning beam at different times and appropriate ranges. The whole receiving system is observed doing its job. However, this technique generates only one threat and is often quite expensive.

Injection Point B. Broadcast Simulator: This technique transmits a signal directly toward a receiver being tested. The transmitted signal will include a simulation of the scanning of the threat antenna. An advantage of this type of simulation is that multiple signals can be transmitted by a single simulator. If a directional antenna (with significant gain) is used, the simulation transmission can be at a relatively low power level, and interference with other receivers will be reduced by the antenna's narrow beamwidth.

Injection Point C. Received Signal Energy Simulator: This technique transmits a signal directly into a receiving antenna, normally using an isolating cap to limit the transmission to the selected antenna. An advantage of this injection point is that the whole receiving system is tested. Coordinated transmissions from multiple caps can be used to test multiple antenna arrays—direction finding arrays, for example.

Injection Point D. RF Signal Simulator: This technique injects a signal that appears to have come from the output of the receiving antenna. It is at the transmitted frequency and at the appropriate signal strength for a signal from the antenna. The amplitude of the signal is modified to simulate the variation in antenna gain as a function of angle of arrival. For multiple-antenna systems, coordinated RF signals are typically injected into each of the RF ports, simulating the cooperative action of the antennas in direction-finding operation.

Injection Point E. IF Signal Simulator: This technique injects a signal into the system at an intermediate frequency (IF). It has the advantage that it does not require a synthesizer to generate the full range of transmission frequencies (since, of course, the system converts all RF inputs to the IF). However, the simulator must sense the tuning controls from the EW system so that it will input an IF signal only when the RF front end of the system (if it were present) would have been tuned to a threat-signal frequency. Any type of modulation can be applied to the IF injected signal. The dynamic range of the signal at the IF input is often reduced from the dynamic range that the RF circuitry must handle.

Injection Point F. Audio- or Video-Input Simulator: This technique is appropriate only if there is some kind of unusual sophistication in the interface between the IF and the audio or video circuitry. Normally, you would chose injection points **E** or **G** rather than this point.

Injection Point G. Audio- or Video-Output Simulator: This very common technique injects audio or video signals into the processor. The injected signals have all of the effects of the upstream path elements, including the effects of any upstream control functions that are initiated by the processor or the operator. Particularly in systems with digitally driven displays, this technique provides excellent realism at the minimum cost. It also allows the checkout of system software at the minimum simulation complexity and cost. It can simulate the presence of many signals at the system antennas.

Injection Point H. Displayed Signal Simulator: This is different from the operator interface simulation we have been talking about in that it injects signals into the actual hardware which displays to the operator. It is appropriate only when analog-display hardware is used. It can test both the operation of the display hardware and the operator's (perhaps sophisticated) operation of that hardware.

11.6.3 General Advantages and Disadvantages of Injection Points

In general, the farther forward in the process a signal is injected, the more realistic the simulation of EW system operation. Care must be taken if receiving equipment anomalies need to be accurately represented in a simulated signal. In general, the farther downstream in the process injection is made, the less complex and expensive the simulation will be. In general, transmitted emulation techniques must be restricted to signals that are unclassified, so real enemy modulations and frequencies may not be used. However, techniques in which signals are hard cabled to the EW system can use real signal characteristics for the most realistic possible testing of software and training of operators.

11.7 Antenna Emulation

In an emulation simulator that inputs signals into a receiver, it is necessary to create the signal characteristics caused by the receiving antenna.

11.7.1 Antenna Characteristics

An antenna is characterized by both gain and directivity. If the receiving antenna is pointed in the direction of an emitter, the signal received from that emitter is increased by the antenna gain, which varies from about -20 dB to $+55$ dB, depending on the type and size of the antenna and the signal frequency. The antenna's directivity is provided by its gain pattern. The gain pattern shows the antenna's gain (usually relative to the gain at its boresight) as a function of the angle from boresight to the direction of arrival (DOA) of the signal.

11.7.2 Simulation of Antenna Functions

In an antenna simulator, the gain at boresight (also called main beam peak gain) is simulated by increasing (or decreasing) the power from the signal generator that is generating the RF signal. Simulating the DOA is a little more complicated.

As shown in Figure 11.24, the DOA of each "received" signal must be programmed into the simulator. In some emulator systems, a single RF generator can simulate several different non-time-coincident emitters. These signals are usually pulsed, but can be any short-duty-cycle signals. In this case, the antenna simulator must be told which emitter it is simulating (with enough lead time to set up its parameters for that signal). The antenna control function is not present in all systems but, when present, usually rotates a single antenna or selects antennas.

For directional antennas (as opposed to those that have fairly constant gain in all directions of interest) the signal from the signal generator is attenuated as a function of the angle from the boresight of the antenna to the simulated DOA of the signal. The antenna orientation is determined by reading the output of the antenna control function.

11.7.3 Parabolic Antenna Example

Figure 11.25 shows the gain pattern of a parabolic antenna. The direction in which the antenna has maximum gain is called the boresight. As the DOA of an emitter moves away from this angle, the antenna gain (applied to that signal) decreases sharply. The gain pattern goes through a null at the edge of the main beam and then forms side lobes. The pattern shown is in a single dimension (e.g., azimuth). There will also be an orthogonal direction pattern (i.e., elevation in this case). This gain pattern is measured by rotating the antenna in an anechoic chamber. The measured gain pattern can be stored in

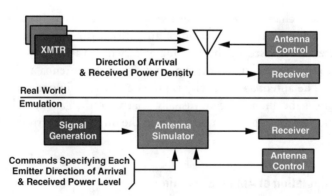

Figure 11.24 The antenna simulator is required to provide an input to the receiver that includes the combined signals from all emitters which the antenna receives.

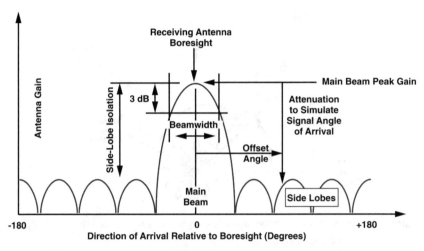

Figure 11.25 In a typical antenna simulator, the direction of arrival of a signal is simulated by adjusting the signal strength to reproduce the antenna gain at the angle of arrival.

a digital file (gain versus angle) and used to determine the attenuation required to simulate any desired angle of arrival.

Although the side lobes in a real antenna will be uneven in amplitude, the side lobes in an antenna simulator are often constant. Their amplitude is lower than the boresight level by an amount equal to the specified side-lobe isolation in the simulated antenna.

If the simulation is to test a receiving system that operates with a rotating parabolic antenna, each of the simulated target signals would be input to the simulator (from the signal generator) at a signal strength which includes the antenna boresight gain. Then, as the antenna control rotates the antenna (either manually or automatically), additional attenuation is added by the antenna simulator. The amount of attenuation is appropriate to simulate the received antenna gain at the offset angle. The offset angle is calculated as shown in Figure 11.26.

11.7.4   RWR Antenna Example

The antennas most commonly used in radar warning receivers (RWRs) have peak gain at boresight, which varies significantly with frequency. However, these antennas are designed for optimum gain pattern. At any operating frequency in their range, their gain slope approximates that shown in Figure 11.27. That is, the gain decreases by a constant amount (in dB) as a

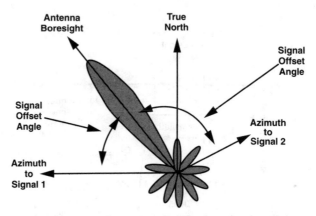

Figure 11.26 Each simulated threat signal is assigned an azimuth, and the offset angle is calculated accordingly.

function of angle from boresight to 90°. Beyond 90°, the gain is negligible (i.e., the signal from the antenna is ignored by the processing for angles beyond 90°).

The complicated part of this is that the antenna gain pattern is conically symmetrical about the boresight. This means that the antenna simulator must create attenuation proportional to the spherical angle between the boresight of the antenna and the DOA of each signal, as shown in Figure 11.28.

These antennas are typically mounted 45° and 135° to the nose of the aircraft and depressed a few degrees below the yaw plane. Add to this the fact that tactical aircraft often do not fly with their wings level and thus may be attacked by threats from any (spherical) angle of arrival.

The typical approach to calculation of the offset angle is to first calculate the azimuth and elevation components of the threat angle of arrival at

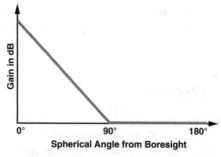

Figure 11.27 The gain pattern of a typical RWR antenna reduces by a fixed amount in dB per degree from boresight—the true spherical angle.

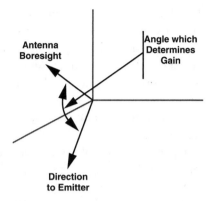

Figure 11.28 The true angle between the antenna boresight and the direction of arrival of the signal depends on the relative location of the transmitter and receiver and the orientation of the vehicle on which the antenna is located.

the aircraft's location—then to set up a spherical triangle to calculate the spherical angle between each individual antenna and the vector toward the emitter.

In a typical RWR-simulator application, there will be one antenna output port for each antenna on the aircraft (four or more). The spherical angle from each threat to the boresight of a single antenna will be calculated for each antenna output and the attenuation set accordingly.

11.7.5 Other Multiple Antenna Simulators

Direction-finding systems that measure the phase difference between signals arriving at two antennas (interferometers) require very complicated simulators—or very simple ones. To provide continuously variable phase relationships is very complex, because the phase measurements are very precise (sometimes part of an electrical degree). For this reason, many systems are tested using sets of cables that have the appropriate length relationships to create the correct phase relationship for a single DOA.

11.8   Receiver Emulation

The previous section discussed antenna emulation. An antenna emulator produces signals for input to a receiver as though the receiver were in some specified operational situation. This section considers how to emulate that receiver.

As shown in Figure 11.29, an RF generator can produce signals that represent transmitted signals arriving at the location of the receiver. An antenna emulator will adjust the signal strength to represent the action of the receiving antenna. The receiver emulator then determines the operator control actions and produces appropriate output signals as though a receiver had been controlled in that manner.

It is generally not useful to emulate just the receiver functions, but rather to include the receiver functions in a simulator (emulator) that represents everything that happens upstream of the receiver output. Such a combined emulator is typically told (through digital inputs) the parameters of received signals and the operator-control actions. In response to that information, the emulator produces appropriate output signals.

11.8.1 The Receiver Function

Our purpose in this section is to consider the portion of that emulation that represents the receiver. First, let's consider the receiver function—apart from the mechanics of receiver design. At its most basic, a receiver is a device that recovers the modulation of a signal arriving at the output of an antenna. To recover that modulation, the receiver must be tuned to the frequency of the signal and must have a discriminator appropriate to the type of modulation on that signal.

A receiver emulator will accept values for the parameters of signals arriving at the receiver input and will read the controls set by an operator. The receiver will then generate output signals representing the outputs that would be present if a particular signal (or signals) were present and the operator had input those control operations.

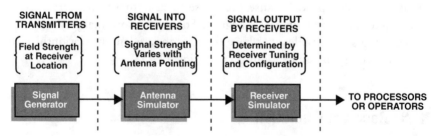

Figure 11.29   The simulation of received signals can be separated into considerations of the intercept geometry, the receiving antenna position, and the configuration of the receiver.

11.8.2 Receiver Signal Flow

Figure 11.30 shows a basic functional block diagram for a typical receiver. It could be in any frequency range and operate against any type of signals.

This receiver has a tuner that includes a tuned preselection filter with a relatively wide pass band. The output of the tuner is a wide-band, intermediate-frequency (WBIF) signal that is output to an intermediate frequency panoramic (pan) display. The IF pan display shows all signals in the preselector pass band. The preselector is usually several MHz wide, and the WBIF is usually centered at one of several standard IF frequencies (455 kHz, 10.7 MHz, 21.4 MHz, 60 MHz, 140 MHz, or 160 MHz, depending on the frequency range of the receiver). Any signal that is received by the tuner will be present in the WBIF output. As the receiver tunes across a signal, the IF pan display will show that signal moving across the tuner pass band in the opposite direction. The center of the WBIF band represents the frequency to which the receiver is tuned, and signals in this output vary with the received signal strength.

The WBIF signal is passed to an IF amplifier which includes (in this receiver) several selectable band-pass filters centered at the IF frequency. This postulated receiver has a narrow-band, intermediate-frequency (NBIF) output—perhaps to drive a direction-finding or predetection-recording function. The NBIF signal has the selected bandwidth. The signal strength of NBIF signals is a function of the received signal strength—but that relationship may not be linear, since the IF amplifier may have a logarithmic response or may include automatic gain control (AGC).

The NBIF signal is passed to one of several discriminators, selected by an operator (or a computer). The demodulated signal is audio or video. Its amplitude and frequency are not dependent on the received signal strength,

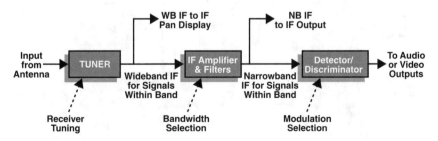

Figure 11.30 This diagram of a typical receiver deals only with the basic receiving functions, independent of the operating frequency and design details.

but rather on the modulation parameters applied to the received signal by the transmitter.

11.8.3 The Emulator

Figure 11.31 shows one way an emulator for this receiver could be implemented. If the emulator were used to test a piece of processing hardware, it would probably be necessary to simulate the anomalies of the receiver. However, if the purpose is to train operators, it will probably be sufficient to just kill the output when the receiver is not properly tuned or the improper discrimination has been selected.

Figure 11.32 shows the frequency and modulation part of the emulator logic for a receiver emulator designed for training. Call the frequency of a simulated signal "SF" and the receiver tuning frequency (i.e., the tuning commands input to the simulator) "RTF."

For a signal to be displayed in the WBIF output, the absolute difference between SF and TRF must be less than half the WB IF bandwidth (WBIF BW). Its output frequency will be:

$$\text{Frequency} = \text{SF} - \text{RTF} + \text{IF}$$

"IF" is the WBIF center frequency. This will cause the generated signal to move across an IF pan display in the opposite direction to the tuning. Note that this signal should have the proper modulation on it.

If the absolute difference between SF and RTF is less than half of the selected NBIF bandwidth, the signal will be present at the NBIF output. Its

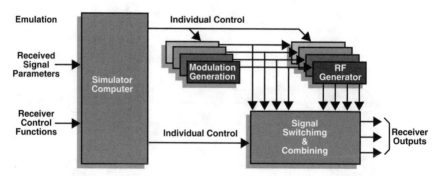

Figure 11.31   The receiver simulator is required to provide signals that would be output from the receiver if the receiver tuning and mode commands are proper for the received signal being simulated.

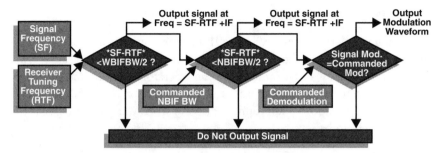

Figure 11.32   The receiver emulator logic determines the output signals as a function of the receiver control inputs from an operator or a control computer.

frequency will be determined by the same equation used for the WBIF output, but now "IF" is the center frequency of the NBIF.

Since this is a training simulator, the logic only requires that the modulation of the received signal match the demodulator that the operator has selected.

11.8.4   Signal-Strength Emulation

The signal-strength input to a receiver depends on the effective radiated power of the signal and the antenna gain at its direction of arrival. As shown in Figure 11.33, the signal strength at each IF output is dependent on the net-gain transfer function. In this receiver, WBIF output level is linearly related to the received signal strength because the gains and losses through the tuner are linear. The NBIF output is proportional to the logarithm of the received signal strength since the IF amplifier has a log transfer function.

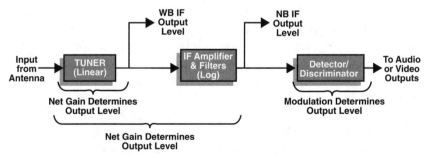

Figure 11.33   The IF output levels are determined by the net gain from the receiver input to the signal output. The level of the modulation determines the audio- or video-output level.

11.8.5 Processor Emulation

In general, modern processors accept IF or video outputs and derive information about received signals (direction of arrival, modulation, etc.). This information is most often presented to an operator as computer-generated audio or visual indications. Thus, the emulation of the processing just involves the generation of appropriate displays when the simulation calls for specific signals to be received.

11.9 Threat Emulation

In Sections 11.7 and 11.8, we have discussed receiver hardware emulation. Now we will talk about the generation of emulated signals.

11.9.1 Types of Threat Emulations

Depending on the point in the receiving system at which the emulation is injected, the signals can be represented by their audio or video modulation, by modulated IF signals, or by modulated RF signals. The following discussion describes the nature of both types of emulated signals and how they are generated.

11.9.2 Pulsed Radar Signals

Modern radars can be pulsed (with or without modulation on the pulses), continuous wave (CW), or continuously modulated. First consider pulsed signals. As shown in Figure 11.34, each signal in the threat environment has its own pulse train. The figure shows a very simple environment including two signals with fixed pulse-repetition intervals (PRI). They are shown with different pulse width and amplitude so they can be easily discriminated. An emulated input to a receiver with enough bandwidth to accept both signals would include the interleaved pulse trains as shown. A realistic environment for a wide-band receiver would include several signals with a total pulse density up to millions of pulses per second.

Next consider the radar's antenna scan characteristics as observed by the receiver. As shown in Figure 11.35, a parabolic antenna has a large main beam and smaller side lobes. As the antenna sweeps past the receiver's location, the threat antenna causes a time varying signal-strength pattern as shown at the bottom of the figure. The elapsed time between receipt of the main lobes is the scan period of the threat antenna. There are many types of

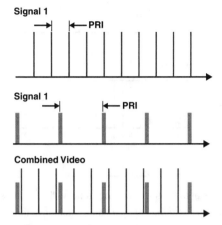

Figure 11.34 The video emulation of a signal environment includes all of the pulses from signals that fall within the receiver's bandwidth.

threat-antenna scans, each of which causes a different received power-versus-time pattern. The next section will consider several typical scans and the way they look to the receiver.

The pulses of a signal with this scan pattern are shown in Figure 11.36. The pulse train (a) is power-adjusted to fit the scan pattern (b), as shown in (c). In line (d), at the bottom of the figure, one of the pulse trains of Figure 11.34 is modified to represent this scanning radar signal. This combined

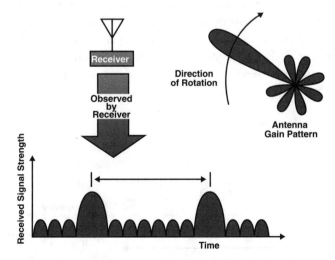

Figure 11.35 The gain pattern of a scanning threat antenna is seen by a receiver as a time-varying signal amplitude.

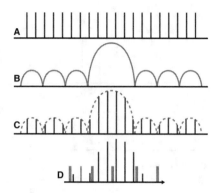

Figure 11.36 A pulsed signal from a scanning radar will have pulse-to-pulse amplitude variation to reflect the changing antenna gain in the direction of the receiver as the antenna scans past.

pulse train could be input to a processor in an EW system to emulate the signal environment as the processor would see it.

To input the environment to a receiver, it is necessary to generate RF pulses at the appropriate frequencies. Figure 11.37 shows (again for a very simple environment) the RF frequency that must be present during the period that these two signals are being emulated. Note that the frequency of signal 1 is always present during the signal 1 pulses and that the frequency of signal 2 is present during its pulses. When there are no pulses present, the frequency is irrelevant, because transmission takes place only during the pulses.

For accurate emulation of IF signals, the two signal frequencies in Figure 11.37 would be within the receiver's IF-band pass. For example, if the IF input at which the signals are injected accepts 160 MHz ± 1 MHz, the RF frequencies of the two signals are 1 MHz apart, and the receiver is tuned to the midpoint between the two signals, the IF injection frequencies would be at 159.5 MHz and 160.5 MHz.

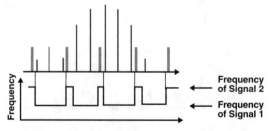

Figure 11.37 For RF emulation of pulsed signals, each pulse needs to be at the correct RF frequency for the signal it represents.

11.9.3 Pulse Signal Emulation

Figure 11.38 shows a basic emulator for multiple pulsed radar signals. This emulator has a number of pulse-scan generators, each creating the pulse and scan characteristics of a single signal. For cost efficiency, the emulator uses one shared RF generator, which must be tuned to the correct RF frequency for each pulse as it is output. This approach is economical because the pulse-scan generators are much less complex than RF generators. Note that the combined pulse-scan outputs can be input to an EW processor, or they can be applied as modulation to the RF generator. However, a synchronization scheme must be applied to tune the RF generator on a pulse-to-pulse basis. If two pulses overlap, the RF generator can only supply the correct frequency for one of the pulses.

11.9.4 Communications Signals

Communications signals have continuous modulation, which carries continually changing information. Signals for audio processing are thus usually supplied from recorder outputs. However, for testing of receivers, it may be practical to use simple-modulation waveforms (sine wave, etc.). When RF-communications signals are emulated, it is necessary to have a separate

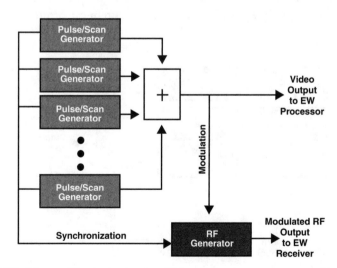

Figure 11.38 The combined outputs of multiple pulse-scan generators can be output to the processor of an EW system, or can be used as the modulation input to an RF generator which is synchronized to provide a combined-signal RF environment to an EW receiver.

RF generator for each signal that is present at any instant. This means that a push-to-talk net (in which only one transmitter is up at a time) can be emulated with a single RF generator and that multiple nets (at different frequencies) can share a single RF generator if the transmissions are short and overlapping signals can be ignored. Otherwise, it's one RF generator per signal.

Figure 11.39 shows the configuration for a typical communications-environment simulator. It should be noted that this same configuration would be used to emulate CW, modulated CW, or pulse-Doppler radars, since each has very high duty factor (or 100%) and thus cannot share an RF generator.

11.9.5 High-Fidelity Pulsed Emulators

Another case in which the dedicated RF-generator configuration is used is that of a high-fidelity pulsed emulator in which no pulse dropouts are acceptable. Since a shared RF generator can only be at one frequency at any given instant, overlapping (or nearly overlapping) pulses require that all but one pulse be dropped. If a processor is processing a pulse train, a missing pulse may cause it to give incorrect answers. The impact of dropped pulses can interfere with effective training or rigorous system testing—so dedicated RF generators are sometimes required where the program can afford them.

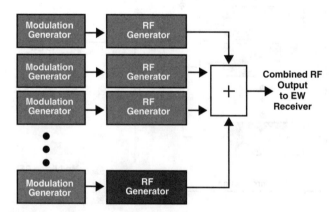

Figure 11.39   Pulsed or communication modulations can be applied to parallel RF generators to generate a communications-signal environment or a radar environment with no interference between pulses from different signals.

11.10 Threat Antenna-Pattern Emulation

The antenna-scanning patterns used by various types of radars depend on their mission. In threat emulation, it is necessary to recreate the time history of the threat antenna gain as seen by a receiver in a fixed location.

The four figures in this section show various types of scans. Each scan type is described in terms of what the antenna is doing and the time history of the threat antenna-gain pattern as it would be seen by a receiver in a fixed location.

11.10.1 Circular Scan

The circularly scanned antenna rotates in a full circle, as shown in Figure 11.40. The received pattern is characterized by even time intervals between observations of the main lobe.

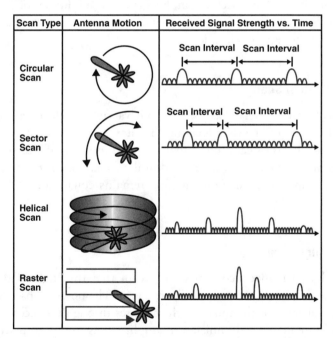

Figure 11.40 Antenna scans classified as circular, sector, helical, and raster are observed by a receiver as very similar. The difference is in the timing and amplitude of the main beam.

11.10.2 Sector Scan

As shown in Figure 11.40, the sector scan differs from the circular scan in that the antenna moves back and forth across a segment of angle. The time interval between main lobes has two values, except in the case in which the receiver is at the center of the scan segment.

11.10.3 Helical Scan

The helical scan covers 360° of azimuth and changes its elevation from scan to scan, as shown in Figure 11.40. It is observed with constant main-lobe time intervals, but the amplitude of the main lobe decreases as the threat antenna elevation moves away from the elevation of the receiver location.

11.10.4 Raster Scan

The raster scan covers an angular area in parallel lines, as shown in Figure 11.40. It is observed as a sector scan, but with the amplitudes of the main-lobe intercepts reduced as the threat antenna covers raster "lines" that do not pass through the receiver's location.

11.10.5 Conical Scan

The conical scan is observed as a sinusoidally varying waveform, as shown in Figure 11.41. As the receiver location (T) moves toward the center of the cone formed by the scanning antenna, the amplitude of the sine wave decreases. When the receiver is centered in the cone, there is no variation in the signal amplitude, since the antenna remains equally offset from the receiver.

11.10.6 Spiral Scan

The spiral scan is like a conical scan, except that the angle of the cone increases or decreases, as shown in Figure 11.41. The observed pattern looks like a conical scan for the rotation which passes through the receiver's location. The antenna gain diminishes in amplitude as the spiral path moves away from the receiver location. The irregularity of this pattern comes from the time history of the angle between the antenna beam and the receiver location.

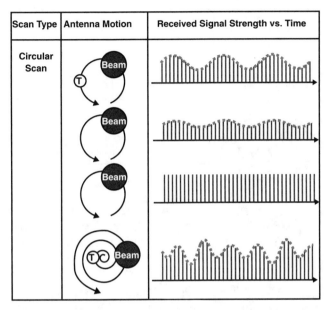

Scan Type	Antenna Motion	Received Signal Strength vs. Time

Figure 11.41 Conical-scan antennas are received with a sinusoidal amplitude pattern. The amplitude of the sinusoid varies with the location of the receiver in the beam. A helical scan is received with a similar pattern, but the apparent location of the receiver in the cone varies as the antenna spirals in or out.

11.10.7 Palmer Scan

The Palmer scan is a circular scan that is moved linearly, as shown in Figure 11.42. If the receiver were right in the middle of one of the circles, the amplitude would be constant for that rotation. In the figure, it is assumed that the receiver is close to the center, but not exactly centered. Therefore, the third cycle shown is a low-amplitude sine wave. As the cone moves away from the receiver location, the sine wave becomes full size, but the amplitude of the signal diminishes.

11.10.8 Palmer-Raster Scan

If the conical scan is moved in a raster pattern, as shown in Figure 11.42, the received-threat gain history will look like the Palmer scan for the line of the raster which moves through the receiver location. Otherwise, the pattern becomes almost sinusoidal, with diminishing amplitude as the raster lines move farther from the angle of the receiver location.

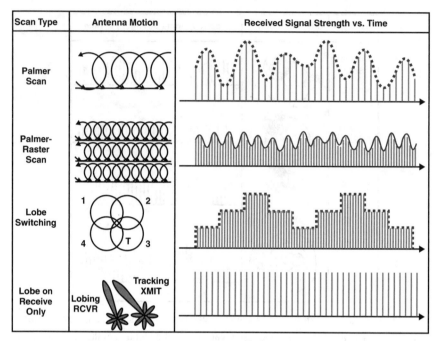

Scan Type	Antenna Motion	Received Signal Strength vs. Time
Palmer Scan		
Palmer-Raster Scan		
Lobe Switching		
Lobe on Receive Only		

Figure 11.42   A Palmer scan is a conical scan that is moved through a linear range. The Palmer-raster is a conical scan that moves in a raster pattern. Lobing antennas show a stepping amplitude pattern as the transmit antenna switches through its lobes. If only the receiving antenna is lobed, the receiver sees a constant amplitude signal.

11.10.9   Lobe Switching

The antenna snaps between four pointing angles, forming a square, as shown in Figure 11.42, to provide the requisite tracking information. Like the other patterns, the received threat antenna-gain history is a function of the angle between the threat antenna and the receiver's location.

11.10.10   Lobe-On Receive Only

In this case, as shown in Figure 11.42, the threat radar tracks the target (the receiver location) and keeps its transmitting antenna pointed at the target. The receiving antenna has lobe switching to provide tracking information. The receiver sees a constant signal level, since the transmit antenna is always pointed at it.

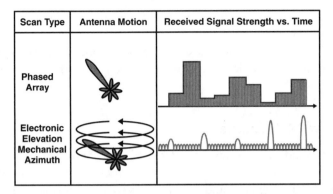

Figure 11.43 A phased-array antenna can be moved directly from any pointing angle to any other pointing angle, so the received pattern will have randomly changing amplitude. If the antenna has a vertical phased array with a mechanical azimuth control, it will look like a circular scan but with a randomly changing main-beam amplitude.

11.10.11 Phased Array

Since a phased array is electronically steered, as shown in Figure 11.43, it can randomly move from any pointing angle to any other pointing angle instantly. Thus, there will be no logical amplitude history observed by the receiver. The received gain depends on the angle between the instantaneous threat-antenna pointing angle and the receiver location.

11.10.12 Electronic Elevation Scan with Mechanical Azimuth Scan

In this case, as shown in Figure 11.43, the threat antenna is assumed to have a circular scan with the elevation arbitrarily moved by a vertical phased array, furnishing a constant time interval between main lobes, but their amplitude can vary without any logical sequence. The azimuth scan can also be a sector scan or can be commanded to fixed azimuths.

11.11 Multiple-Signal Emulation

A characteristic of an EW threat environment is that many of the signals have short duty cycles. Therefore, it is possible to use a single generator to produce more than one threat signal. This has the advantage of significantly decreasing the cost per signal. However, as you will see, this cost saving can come at a performance cost. In this section we will discuss the various ways to achieve multiple-signal emulation.

The following discussion covers two basic methods for the emulation of multiple signals. The basic tradeoff between the two methods is cost versus fidelity.

11.11.1 Parallel Generators

For maximum fidelity, a simulator is designed with complete parallel simulation channels, as shown in Figure 11.44. Each channel has a modulation generator, an RF generator, and an attenuator. The attenuator can simulate the threat scan and the range loss, as well as (if appropriate) the receiving-antenna pattern. The modulation generator can provide any type of threat modulation: pulse, CW, or modulated CW. This configuration can provide more signals than the number of channels, since not all signals will be simultaneous. However, it can provide a number of *instantaneously simultaneous* signals equal to the number of channels. For example, with four channels, it could provide a CW signal and three overlapping pulses.

11.11.2 Time-Shared Generators

If only one signal needs to be present at any instant, a single set of simulation components (as shown in Figure 11.45) can provide many signals. This con-

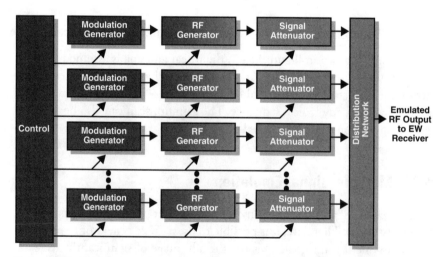

Figure 11.44 The combined outputs of multiple simulation channels can be combined to generate a very accurate representation of complex signal environments.

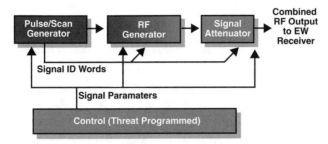

Figure 11.45 A single set of emulation components can be used to provide a multiple-signal-pulse output by controlling each component on a pulse-to-pulse basis.

figuration is normally used only for a pulsed-signal environment. A control subsystem contains the timing and parametric information for all of the signals to be emulated. It controls each of the simulation components on a pulse-to-pulse basis. The drawback of this approach is that it can only output one RF signal at any given instant. This means that it could output one CW or modulated CW signal, or any number of pulsed signals as long as their pulses do not overlap. As shown below, there is actually a restriction on pulses that closely approach each other in time, even if they do not actually overlap.

11.11.3 A Simple Pulse-Signal Scenario

Figure 11.46 shows a very simple pulse scenario with three signals that contain no overlapping pulses. All of these pulses can be supplied by a single simulator string controlled on a pulse-by-pulse basis. Figure 11.47 shows the combined video from the three signals on the first line. This is the signal that would be received by a crystal video receiver covering the frequencies of all

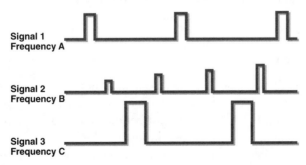

Figure 11.46 This is a very simple pulse scenario with three signals. In this example, the pulses do not overlap.

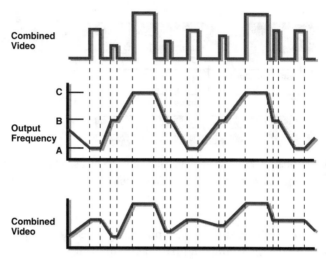

Figure 11.47   Combining the three signals and controlling the simulator on a pulse-by-pulse basis require these changes in output frequency and power.

three signals. The second line shows the frequency control that would be required to include all three signals in the RF output from the simulator. Note that the correct signal frequency must remain for the full pulse duration. Then, the synthesizer in the RF simulator has the interpulse time to tune to the frequency of the next pulse. Note that the synthesizer tuning and settling speed must be fast enough to change by the full frequency range in the shortest specified interpulse period. The third line of the figure shows the output power required to simulate all signals on a pulse-by-pulse basis. This means that the attenuator must settle at the correct level with the required accuracy during the minimum interpulse time. The change between pulses can be up to the full attenuation range. Depending on the simulator configuration, this attenuation can be just for the threat scan and range attenuation or can include receiving-antenna simulation.

11.11.4   Pulse Dropout

Before the synthesizer and attenuator can start moving to the proper values for the next pulse, they must receive a control signal. This control signal, a digital word ("signal-ID word"), is sent before the leading edge of the pulse by an "anticipation time," as shown in Figure 11.48. The anticipation time must be long enough for both the worst-case attenuator-settling time and the worst-case synthesizer-settling time. The longer of the two times dictates the anticipation time. In the figure, the worst-case attenuator-settling time is

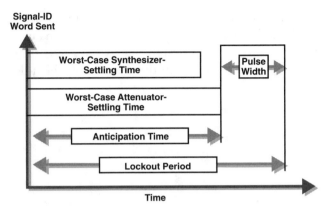

Figure 11.48 The control signal must anticipate each pulse by sufficient time to allow both the frequency and output-power settings to stabalize. Subsequent pulses are locked out by the sum of the anticipation time and the pulse width.

shown as longer than the worst-case synthesizer-settling time. The "lockout period" is the time delay after the signal-ID word before another signal-ID word can be sent. If a pulse occurs within a time that is the sum of the pulse width of the previous pulse and the anticipation time, it will be dropped from the simulator output.

11.11.5 Primary and Backup Simulators

The percentage of pulse dropouts can be significantly reduced if a second simulator channel is used to provide pulses that are dropped by the primary simulator channel. The analysis of the percentage of pulses dropped in various simulator configurations uses the binomial equation, which is also useful in several other EW applications.

11.11.6 Approach Selection

The selection of the approach to providing multiple-signal emulation is a matter of cost versus fidelity. In systems requiring high fidelity and few signals, the choice is clearly to provide full parallel channels. Where slightly lower fidelity (perhaps 1% or 0.1% pulse dropout) is acceptable and there are many signals in the scenario, it may be best to provide a primary simulator with a one or more secondary simulators. Where pulse dropouts can be tolerated, a single-channel simulator will do the job at significant cost saving. By setting priorities among signals, so as to avoid dropping pulses from higher-priority signals, the impact of dropped pulses may be minimized.

One approach that may give excellent results is to use dedicated simulators for specified threat emitters, while using a single-channel, multiple-signal generator to provide background signals. This tests the ability of a system to process a specified signal in the presence of a high-density pulse environment.

Appendix A:

Cross-Reference to EW 101 Columns in the *Journal of Electronic Defense*

Chapter 1	No columns included
Chapter 2	July and September 1995; and March and April 2000 columns
Chapter 3	September, October, November, and December 1997 columns
Chapter 4	August, October, and December 1995; and January and April 1996 columns
Chapter 5	October, November, and December 1998; and January, February, and March 1999 columns
Chapter 6	January, February, March, April, and part of May 1998 columns
Chapter 7	Part of May 1998; and June, July, August, and September 1998 columns
Chapter 8	October 1994; and January, February, March, April, May, and June 1995 columns
Chapter 9	May, June, July, August, November, and December 1996; and January, February, March, and April 1997 columns
Chapter 10	May, June, July, and August 1997 columns
Chapter 11	April, May, June, July, August, September, October, November, and December 1999; and January and February 2000 columns

About the Author

David Adamy is an internationally recognized expert in electronic warfare . . . perhaps mainly because he has been writing the EW 101 columns for many years. In addition to writing the columns, however, he has been an EW professional (proudly calling himself a "Crow") in and out of uniform for 38 years. As a systems engineer, project leader, program technical director, program manager, and line manager, he has directly participated in EW programs from just above DC to just above light. Those programs have produced systems that were deployed on platforms ranging from submarines to space and that met requirements from "quick and dirty" to high reliability.

He holds BSEE and MSEE degrees, both in communication theory. In addition to the EW 101 columns, he has published many technical articles in EW, reconnaissance, and related fields and has seven books in print (including this one). He teaches EW-related courses all over the world and consults for military agencies and EW companies. He is a long-time member of the National Board of Directors of the Association of Old Crows, and he runs the organization's professional development courses and the technical tracks at its annual technical symposium.

He has been married to the same long-suffering wife for 40 years (she deserves a medal for putting up with a classical nerd that long) and has four daughters and six grandchildren. He claims to be an OK engineer but one of the world's truly great fly fishermen.

Index

Radar Evaluation Handbook, David K. Barton, et al.

Radar Meteorology, Henri Sauvageot

Radar Signal Processing and Adaptive Systems, Ramon Nitzberg

Radar System Performance Modeling, G. Richard Curry

Radar Technology Encyclopedia, David K. Barton and Sergey A. Leonov, editors

Range-Doppler Radar Imaging and Motion Compensation, Jae Sok Son, et al.

Theory and Practice of Radar Target Identification, August W. Rihaczek and Stephen J. Hershkowitz

For further information on these and other Artech House titles, including previously considered out-of-print books now available through our In-Print-Forever® (IPF®) program, contact:

Artech House	Artech House
685 Canton Street	46 Gillingham Street
Norwood, MA 02062	London SW1V 1AH UK
Phone: 781-769-9750	Phone: +44 (0)20 7596-8750
Fax: 781-769-6334	Fax: +44 (0)20 7630-0166
e-mail: artech@artechhouse.com	e-mail: artech-uk@artechhouse.com

Find us on the World Wide Web at:
www.artechhouse.com